Student Study Guide with Selected Solutions for
University Chemistry

Joseph Noroski
University of Pittsburgh

Peter Siska
University of Pittsburgh

PEARSON

Benjamin
Cummings

San Francisco   Boston   New York
Cape Town   Hong Kong   London   Madrid   Mexico City
Montreal   Munich   Paris   Singapore   Sydney   Tokyo   Toronto

Publisher *Jim Smith*
Marketing Manager *Scott Dustan*
Project Editors *Katherine Brayton, Katie Conley*
Assistant Editor *Cinnamon Hearst*
Editorial Assistant *Kristin Rose*
Production Supervisor *Shannon Tozier*
Manufacturing Buyer *Carol Melville*
Cover Designer *Jeff Puda, Karl Miyajima*
Text and Cover Printer *Bradford & Bigelow*

ISBN 0-8053-3085-2

1 2 3 4 5 6 7 8 9 10—B&B—08 07 06 05
www.aw-bc.com

TO THE STUDENT

Using a more comprehensive presentation than what is encountered in a typical general chemistry textbook, *University Chemistry* has been written to challenge advanced freshmen with a solid background in chemistry. This goal can't be achieved by providing stale Exercises with virtually rubber stamp solutions. Accordingly, the Exercises that appear at the end of each chapter are designed to *increase* your knowledge and *empower* you to tackle new aspects of chemistry that you have never seen. Each Exercise has been written with you and your future in mind. *University Chemistry* was written for you. True learning has a price, however. It takes time and patience. As a freshman science or engineering major, or just a student interested in chemistry, you are only beginning. You can't hope to learn the all of the material in *University Chemistry* in one or two college terms. Nothing prevents you from learning this material if you persevere. Do not give up.

Specifically, this Student Solutions Manual contains, for each Chapter, a list of goals that you are expected to accomplish, a list of key equations, an overview of selected concepts, tips for solving many of the even numbered Exercises, and complete solutions to the odd numbered Exercises. Additionally, the introductions to each Chapter frequently highlight recurring concepts by making references to various Exercises, Equations, Figures, etc. throughout *University Chemistry*, not just those within a particular Chapter. To avoid redundancy, we have not included a vocabulary list here, but the Summary of each Chapter in *University Chemistry* contains boldface words. These can be regarded as the key terms or vocabulary list for each chapter. A word of caution. You will learn very little and retain even less if you read these solutions without giving serious effort to the Exercises first. In fact there is a good story to support this position. The proofreading of the Exercises for *University Chemistry* was done independently of the author. The proofreader did not have access to a "solutions manual". Such a process of reading the answers, which is always accompanied by the requisite head nodding in approval and responses of, 'Oh, yes. Sure. Of course!', would have been pointless. It was only when the author and proofreader *independently* came to agreement on the answer that the Exercise was considered to be solved. In this manner inconsistencies were removed. We urge you to take this approach. *Verify* our work. Don't just accept it!

1

Physical Principles
Underlying Chemistry

1 Physical Principles Underlying Chemistry

Your Chapter 1 GOALS:

- Understand the meaning and importance of the Laws of Definite Proportions, Multiple Proportions, and Combining Volumes
- Learn Dmitri Mendeleyev's Periodic Law and his influence on the Periodic Table
- See that physics is an indispensable tool when describing chemical phenomena
- Define and grasp the relationship between work, kinetic energy, and potential energy
- Express temperatures in the Fahrenheit, Celsius, and Kelvin scales
- Describe the structure of an atom
- Define isotopes and use their masses to calculate the weighted average atomic mass that appears on the Periodic Table
- Learn Coulomb's Law and how it is used to derive Equation 1.18
- Become fluent in unit conversion between cgs and mks unit systems
- Be able to solve stoichiometry problems in all of their guises: titration, empirical formula, molecular formula, percent composition, and aqueous solution problems

Chapter 1 KEY EQUATIONS:

- 1.9, 1.10, 1.11, 1.12, 1.13, 1.15, 1.17, 1.18
- $V(eV) = 14.40 q_1 q_2 / r(\text{Å})$
- Potential energy expressions in Figure 1.9
- $L = mvr$ (See Exercise 13.)

Overview

Parts of Chapter 1 ought to be quite familiar to you! Namely, the brief historical outline detailing the elucidation of the atom's structure, SI units, (the sometimes maddening) significant figures, Mendeleyev's Periodic Law, the atom's constituents, and many others. In fact, your likely ease in understanding Example 1.6 is alone proof of your mastery of basic chemical principles. Be proud of the knowledge you already possess. Coupled with this familiar material, however, is a bevy of additional information that must also become a part of your chemical foundation.

Oddly enough, if we had to describe this (probably) new material in one word, it would be…*physics*. What? This is chemistry! Why all the physics? The answer is a fundamental one, and it can be summed up in two parts, based on the two most important things to chemists: atoms and molecules. Atoms contain protons and electrons, which are both charged. (Obviously, the neutrons are of interest to chemists in mass and isotope considerations.) The primary (classical) force that governs the interactions of charged particles is described by Coulomb's Law. (See Section 1.7.) As for molecules, they are, of course, made up of atoms. The (classical) force that governs the motion, or vibrations, of atoms *in bonds* can be modeled after (that is, 'assumed to be similar to', or 'can be approximated by') Hooke's Law. (See Example 1.4.) Therefore, any attempt to describe the building blocks of chemists, atoms and molecules, inherently relies on laws that are traditionally attributed to the field of physics. Using physics is simply inescapable if we chemists are to have any quantitative description of matter. Ultimately, this implies that the lines between scientific disciplines are arbitrary, and we encourage you to view the natural sciences as a whole.

The phrase "quantitative description" leads us to another, perhaps, unfamiliar area – cgs units. It is a grim truth that, even though a single unit system would simplify science, scientists simply can't agree on which system is best. Some use only SI units, others use cgs units, and others use British units. As a student of science, you have no choice but to become fluent in (at least) the mks and cgs systems. One advantage of the cgs system is particularly evident in Equations 1.17 and 1.18, where $k = 1$, not $1/4\pi\varepsilon_0$ as in the mks system. (See Section 1.7 and Exercise 40.) The good news is that no matter what unit system you use, you must get the right unit in your answer as long as your units are consistent. For example, in the cgs system, all force equations must come out in dynes. The most common properties and their mks and cgs names are compiled below:

property	mks name(symbol)	conversion factor	cgs name
force	Newton (N)	10^5	dyne
energy	Joule (J)	10^7	erg
charge	Coulomb (C)	2.9979×10^9	esu

Multiplying the mks value by the given conversion factor yields the cgs value.

While in this physics mood, let us repeat the important concepts revealed by the solution of Exercises 9 through 12. Potential energy is the energy that a body acquires by virtue of its position in a field of force. The force can arise from gravity, separated charges, or compressed springs. In any case it is energy that is stored, waiting to be released. When released, it is converted into kinetic energy. Equations 1.11 and Example 1.4 provide the necessary formulas to solve these problems. Note the give-and-take nature of the solutions. All the energy that is potential is converted to kinetic energy. If the motion is harmonic (Exercise 12), all of that kinetic energy is again returned to potential energy. Common sense tells us that the kinetic energy that is gained can never exceed the potential energy that was there when the body was at rest. There is, however, a hint in Figure 1.10 that what is "impossible in Newton's mechanics" might just not be impossible for electrons. See the discussion of RDF's in Section 3.4.

Throughout this book, we develop "engineering formulas" that can be of great time saving value. These formulas compile all of the constants and conversion factors of an equation into one number, leaving only the variables to insert. In Section 1.7 we derive such a formula for the potential energy V (in eV) of Equation 1.18: $V(eV) = 14.40q_1q_2/r(Å)$. Now, when we say in the text that the charges are in units of e, we refer to the fact that a single proton possesses one unit of charge – the fundamental unit of charge, 1.602×10^{-19} J. The electron possesses a charge that is equal in magnitude but opposite in sign to that of the proton. For example, a lithium nucleus has a charge of $+3e$, and the "q" needed for the engineering formula is 3. Likewise, a sulfide ion has a charge of $-2e$, and the needed "q" is -2. The signs matter, so don't forget them! See Exercises 40 through 43 and 45.

Exercises 7, 19, and 27 explore those pesky fractions. In chemistry, however, the phrase "mole fraction" is often encountered. Don't let the word mole confuse you. You already know that the sum of the fractions of a whole, be they pie pieces or "mole pieces", is one. Thus, when there are only two components, a and b, of a whole, (e.g., two isotopes of an element, or two components in a gas sample), we can express their parts of the whole as X and $(1 - X)$. X is, for example, the fraction of the total number of moles that is a, the mole fraction $n_a/(n_a + n_b)$. $(1 - X)$ is the mole fraction of b, $n_b/(n_a + n_b)$. Further, note that in chemistry problems these fractions are typically "weighted" by the mass of the individual component. See Example 1.5 and the chlorine example before it. Exercise 27(b) is particularly noteworthy for two reasons. First, the phrase "air is 78.084 mol % N_2" simply means that 78.084 % (as a fraction, 0.78084) of air is N_2. This applies to measuring units such as liters, gallons, and moles. For example, if you have 100 L of air, you have 78.084 L of N_2. Note, however, that 0.78084 is *not* a mass fraction. If you have 100 g of air, you do *not* have 78.084 g of N_2. Second, the solution to this Exercise uses a *ratio*. The more familiar phrase to you might be a "conversion factor". That is, the ratio 0.934 L Ar/78.084 L N_2 applies to any air sample. In fact, you are quite familiar with ratios! Simple applications of Boyle's Law and Charles's Law use nothing but ratios. For example, when we solve $P_1 = (V_2/V_1)P_2$, note that P_1 is related to P_2 by the *ratio* V_2/V_1.

Exercise 15 is numerically simple, yet it is, perhaps, the most profound Exercise in Chapter 1. All of thermodynamics is based on the concept of temperature and what temperature reveals about the average energy of the constituents of matter, atoms. For now, settle for becoming familiar with the size and units of Boltzmann's constant and the fact that $\bar{\varepsilon}$ is an average energy (kinetic plus potential) of an individual molecule. The relevance of the size of $k_B T$ will be examined further in Exercise 23 of Chapter 2, Exercise 22 of Chapter 3, Exercise 21 of Chapter 8, and in Chapter 10.

1. Lavoisier's masses, together with his Law of Mass Conservation, allow us to find the mass of oxygen liberated by heating the red calx by difference,

$$m(O) = m(\text{red calx}) - m(Hg) = 2.39 - 2.20 = 0.19 \text{ g O}$$

Note that one significant figure (1S) has been lost in the subtraction. To find the formula for the red calx, the oxide of mercury, we need to know how many atoms N, or moles of atoms n, of Hg and O are present; the formula is then determined by the ratio of these numbers of moles. If $x = n_{Hg}/n_O$, the formula can then be written as Hg_xO. Moles of Hg and O can be found by dividing their respective masses in g by the (modern) atomic masses in g/mol:

$$n_O = (0.19 \text{ g O})\left(\frac{1 \text{ mol O}}{16.00 \text{ g O}}\right) = 0.0119 \text{ mol O}$$

$$n_{Hg} = (2.20 \text{ g Hg})\left(\frac{1 \text{ mol Hg}}{200.6 \text{ g Hg}}\right) = 0.01097 \text{ mol Hg}$$

These yield

$$x = \frac{n_{Hg}}{n_O} = \frac{0.01097 \text{ mol O}}{0.0119 \text{ mol Hg}} = 0.92$$

and the formula $Hg_{0.92}O$. Note that we retained one more S in calculating the n's than the precision of the data warranted; this was done to avoid rounding errors in the final mole ratio. Even better in this regard is to store the "exact" intermediate numbers in your calculator and perform the final ratio calculation using the stored numbers. Dalton's atomic hypothesis suggests that the formula is, within experimental error, simply HgO, which is the accepted modern formula for mercuric oxide.

3. The mass ratios are independent of the mass units (here "grains"), and are simply calculated:

Azotic gas/oxygen gas:	0.30064 / 0.34211 = 0.87878
Hydrogen gas/oxygen gas:	0.02394 / 0.34211 = 0.06998
Carbonic acid gas/oxygen gas:	0.44108 / 0.34211 = 1.2893

According to Avogadro, each "cubical inch" contains the same number of molecules, regardless of their chemical identity, and therefore the mass ratios are the ratios of molecular masses. Using the modern values, we find

N_2 / O_2 :	28.0134 / 31.9988 = 0.87545
H_2 / O_2 :	2.0158 / 31.9988 = 0.06300
CO_2 / O_2 :	44.0098 / 31.9988 = 1.3754

remarkable agreement considering the time Lavoisier's work was done.

5. The combining masses are calculated from modern atomic masses as follows:

$$\text{For } CH_4: \left(\frac{m_C}{m_H}\right)_{ch} = \frac{12.011}{4(1.0079)} = 2.979$$

$$\text{For } C_2H_4: \left(\frac{m_C}{m_H}\right)_{og} = \frac{2(12.011)}{4(1.0079)} = 5.958$$

$$\frac{(m_C/m_H)_{og}}{(m_C/m_H)_{ch}} = \frac{5.958}{2.979} = 2.000$$

Since the ratio of ratios is 2.000 or 2:1, Dalton's Law of Multiple Proportions is satsified; the 2 comes from the numerators of the combining mass ratios, and is just a result of counting the C atoms.

7. If the entire error in the H_2 / O_2 mass ratio is due to air-contamination of the hydrogen sample,

$$\left(\frac{M_{H_2}}{M_{O_2}}\right)_{observed} = r = \frac{(1-X)M_{H_2} + XM_{air}}{M_{O_2}}$$

where r is the observed mass ratio and X is the mole fraction of air in the H_2 sample. Solving for X,

$$X = \frac{rM_{O_2} - M_{H_2}}{M_{air} - M_{H_2}} = \frac{(0.06998)(32.00) - 2.016}{29.0 - 2.0} = 0.00827$$

The sample of H_2 may have contained 0.827 % air, a low level of contamination.

9. First, convert the speed v to m s^{-1}:

$$v(\text{ms}^{-1}) = \frac{55\,\text{mi}}{h}\left(\frac{5280\,\text{ft}}{\text{mi}}\right)\left(\frac{12\,\text{in}}{\text{ft}}\right)\left(\frac{2.54\,\text{cm}}{\text{in}}\right)\left(\frac{1\,\text{m}}{100\,\text{cm}}\right)\left(\frac{1\,\text{h}}{3600\,\text{s}}\right) = 24.6\,\text{ms}^{-1}$$

In this units conversion, all the conversion factors are exact, by definition, and the result can be computed and stored once and for all to produce the "engineering formula" v(m/s) = 0.44704 v(mi/h). The kinetic energy is then

$$K = \tfrac{1}{2}mv^2 = \tfrac{1}{2}\left[2600\,\text{lb}\left(\frac{0.4536\,\text{kg}}{\text{lb}}\right)\right](24.6\,\text{m/s})^2 = 3.6 \times 10^5\,\text{J} = 360\,\text{kJ}$$

The mass of hydrogen gas required to provide this amount of energy is

$$g\,H_2 = 360\,\text{kJ}\left(\frac{1\,\text{mol}\,H_2}{237\,\text{kJ}}\right)\left(\frac{2.016\,\text{g}\,H_2}{\text{mol}}\right) = 3.1\,\text{g}$$

a remarkably tiny fuel supply. The work w done on a frictionless car is numerically equal to K, w = 360 kJ.

11. The potential energy is

$$V(x) = mgx$$
$$= (1.02\,\text{kg})(9.8\,\text{m s}^{-2})x$$
$$= 10x, \; x \text{ in m}$$

At $x = h = 0.100$ m, $K = 0$ and

$E = V = 10(0.100) = 1.0$ J

See graph at the right.

a. $K(x) = E - V(x) = 1.0$ J $- 10x$
   At $x = 0.050$ m, $K_{50} = 0.5$ J
   At $x = 0.000$ m, $K_0 = 1.0$ J

b. $v = [2K/m]^{\frac{1}{2}}$
   $v_{50} = [2(0.5)/1.02]^{\frac{1}{2}} = 0.99$ m/s
   $\qquad = 99$ cm/s
   $v_0 = 140$ cm/s

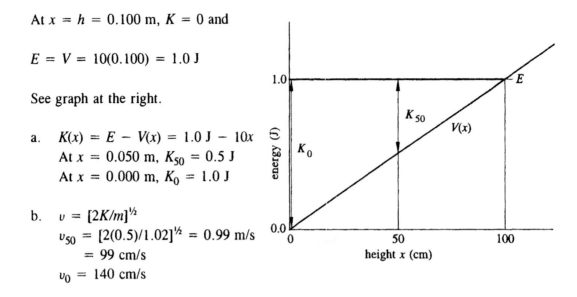

13. The orbital velocity v is also known as the tangential velocity, i.e., the velocity is a vector tangent to the circle through which the ball rotates. Using the hint, we derive

$$L = mr^2 \frac{d\theta}{dt} = mr^2\omega = mr^2 \frac{2\pi}{\tau} = mr^2 \frac{2\pi v}{2\pi r} = mvr$$

The sequence implies that $\omega = v / r$, or $v = r\omega$, allowing the ready conversion between orbital velocity v(m/s) and angular ω(radians/s) for rotating objects. From $L=mvr$ we can determine the units of angular momentum,

$$\text{kg m}^2 / \text{s} = [\text{kg m}^2 / \text{s}^2] \text{ s} = \text{J s}$$

The kinetic energy is then

$$K = \tfrac{1}{2}mv^2 = \tfrac{1}{2}m\left(\frac{L}{mr}\right)^2 = \frac{L^2}{2mr^2}$$

15. Using Equation 1.15,

$$\bar{\varepsilon} \approx k_B T = (1.381 \times 10^{-23} \text{ J/K})(298 \text{ K}) = 4.11 \times 10^{-21} \text{ J}$$

The energy E that one mole of hydrogen molecules possesses is

$$E = \bar{\varepsilon} N_A = (6.022 \times 10^{23} \text{ /mol})(4.11 \times 10^{-21} \text{ J}) = 2480 \text{ J/mol}$$

or, in calories, using Equation 1.16,

$$E = (2480 \text{ J/mol})(1 \text{ cal} / 4.184 \text{ J}) = 592 \text{ cal/mol}$$

17. It is a good approximation at low Z, but not at high Z. As Z increases, the number of neutrons that are needed to create a stable nucleus increases at a faster rate than the number of protons being added. The nucleus will be examined in more detail in Chapter 16.

19. When only two isotopes contribute to the mass of an element, the isotope and elemental masses uniquely determine the percent composition. If we let X be the mole fraction of the lighter (85) isotope of Rb, we have

$$M_{Rb} = X M_{85} + (1 - X) M_{87}$$

Solving for X,

$$X = \frac{M_{87} - M_{Rb}}{M_{87} - M_{85}} = \frac{87 - 85.4678}{87 - 85} = 0.7661$$

We have assumed as suggested that the isotope masses are equal to the mass numbers. Judging from the data for Br in Table 1-1, (or using the table of the isotopes in the Handbook of Chemistry and Physics) this assumption is good to ~0.1 amu; we thus expect the results to be valid to less than 2 significant figures. Rounding, we find that natural Rb consists of 77% ^{85}Rb and 23% ^{87}Rb. (Literature values are 72.2% and 27.8%.)

21. With only two isotopes, there can only be three types of Cl_2 molecules: ^{35}Cl—^{35}Cl ; ^{35}Cl—^{37}Cl ; ^{37}Cl—^{37}Cl. The line represents a bond between the two isotopes of this diatomic molecule. Thus, the peak at 70 is from the first combination, and so on. To get the relative numbers of these three combinations, we need to consider probabilities. Imagine we have two boxes of Cl atoms, each containing the correct isotopic abundance. If you were to reach into the first box, you have a 75.77 % chance of pulling out a ^{35}Cl atom. You have the same chance when you reach into the second box.

Now, the probability of drawing *two* ^{35}Cl's is just the product of the individual probabilities. Consider flipping a coin. The probability of tails is 0.5 for the first flip. The probability of getting tails again is $(0.5)(0.5) = 0.25$. That is, you multiply the probability of each step. Another analogy is guessing your way through a true / false test. If there were 10 questions on the test, your chance of guessing all of them correctly is $(\frac{1}{2})^{10} = 1 / 1024$. These considerations will play a role in Chapter 11.

The probability of getting a ^{35}Cl from the first box and the second box is thus the product of their fractional abundances: $P(35,35) = (0.7577)(0.7577) = 0.5741$. The probability of getting two consecutive ^{37}Cl atoms is $P(37,37) = (0.2423)(0.2423) = 0.05871$. The combination ^{35}Cl and ^{37}Cl, however, can occur in either order; but the order is not important, so you must add $P(35,37)$ and $P(37,35)$. Since these two are equal, the probability is twice $P(35,37) = 2(0.7577)(0.2423) = 0.3672$. It should be noted that the probabilities must add up to one. If we had not counted the 35/37 combination twice, we could have realized that something was wrong since $0.5441 + 0.0587 + 0.1836 < 1$. The probabilities we got do add up to 1, and we predict the following relative heights for the peaks:

$$70 : 72 : 74 = 0.5741 : 0.3672 : 0.0587.$$

23. The principle here is to convert grams to moles to molecules:

$$\text{pure C: atoms C} = 6\,\text{g C}\left(\frac{1\,\text{mol C}}{12.0\,\text{g C}}\right)\left(\frac{6.02 \times 10^{23}\,\text{atoms C}}{1\,\text{mol C}}\right) = 3.0 \times 10^{23}\,\text{atoms}$$

$$CO_2: \text{ atoms C} = 6\,\text{g CO}_2\left(\frac{1\,\text{mol CO}_2}{44.0\,\text{g CO}_2}\right)\left(\frac{1\,\text{mol C}}{1\,\text{mol CO}_2}\right)\left(\frac{6.02 \times 10^{23}\,\text{atoms C}}{1\,\text{mol C}}\right)$$

$$= 0.8 \times 10^{23}\,\text{atoms}$$

$$C_{12}H_{22}O_{11}: \text{ atoms C} = 6\,\text{g C}_{12}H_{22}O_{11}\left(\frac{1\,\text{mol C}_{12}H_{22}O_{11}}{342.3\,\text{g C}_{12}H_{22}O_{11}}\right)\left(\frac{12\,\text{mol C}}{1\,\text{mol C}_{12}H_{22}O_{11}}\right)$$

$$\times\left(\frac{6.02 \times 10^{23}\,\text{atoms C}}{1\,\text{mol C}}\right) = 1.3 \times 10^{23}\,\text{atoms}$$

25. Assuming 100 g elemental samples, we find moles and their ratio,

$$\text{mol Al} = 100.\,\text{g Al}\left(\frac{1\,\text{mol Al}}{26.98\,\text{g Al}}\right) = 3.71\,\text{mol}$$

$$\text{mol O} = 100.\,\text{g O}\left(\frac{1\,\text{mol O}}{16.00\,\text{g O}}\right) = 6.25\,\text{mol}$$

$$\frac{\text{mol O}}{\text{mol Al}} = \frac{6.25}{3.71} = 1.68$$

Since the computed mole ratio O:Al exceeds that in the compound, $3/2 = 1.50$, we conclude that O atoms are in excess, and Al is limiting. Another approach is to divide the computed number of moles by the subscript for each element; the smallest such ratio is the limiting reagent. We then have

% O remaining = 100% − % O reacted

$$= 100\%\left[1 - 3.71\,\text{mol Al}\left(\frac{3\,\text{mol O}}{2\,\text{mol Al}}\right)/(6.25\,\text{mol O})\right] = 11.0\%$$

27. a. We use the reaction stoichiometry to relate moles of Mg to moles of N_2, and then convert to volume using the STP molar volume:

$$n_{Mg} = 18.2\,\text{g}\left(\frac{1\,\text{mol}}{24.305\,\text{g}}\right) = 0.749\,\text{mol Mg}$$

$$V_{N_2} = 0.749\,\text{mol Mg}\left(\frac{1\,\text{mol N}_2}{3\,\text{mol Mg}}\right)(22.4\,\text{L/mol}) = 5.59\,\text{L}$$

b. Invoking Avogadro's principle that number of moles, or mole %, is proportional to volume, we find

$$V_{Ar} = 5.59\,\text{L N}_2\left(\frac{0.934\,\text{L Ar}}{78.084\,\text{L N}_2}\right) = 0.0669\,\text{L Ar} = 66.9\,\text{mL Ar}$$

29. We exploit the fact that all the carbon appears in CO_2 and all the hydrogen in H_2O to find

$$\text{mol C} = 1.637 \text{ g } CO_2\left(\frac{1 \text{ mol } CO_2}{44.01 \text{ g } CO_2}\right)\left(\frac{1 \text{ mol C}}{1 \text{ mol } CO_2}\right) = 0.03720 \text{ mol}$$

$$\text{mol H} = 0.335 \text{ g } H_2O\left(\frac{1 \text{ mol } H_2O}{18.02 \text{ g } H_2O}\right)\left(\frac{2 \text{ mol H}}{1 \text{ mol } H_2O}\right) = 0.0372 \text{ mol}$$

$$\frac{\text{mol H}}{\text{mol C}} = \frac{0.0372}{0.0372} = 1.00$$

The empirical formula is thus CH. We expect the molecular formula to be $(CH)_n$, where n is a positive integer. The balanced combustion reaction in this case would be

$$(CH)_n + \tfrac{5}{4}nO_2 \rightarrow nCO_2 + \tfrac{n}{2}H_2O$$

implying that $n = \text{mol } CO_2 \text{ / mol } (CH)_n$, or

$$n = \frac{0.0372 \text{ mol } CO_2}{0.417 \text{ L } (CH)_n}\left(\frac{22.4 \text{ L } (CH)_n}{1 \text{ mol } (CH)_n}\right) = 2.00$$

The molecular formula is thus $(CH)_2$ or C_2H_2 (acetylene or, properly, ethyne). Can you write the balanced combustion equation?

31. In each case, the moles required $= (0.500 \text{ mol/L})(0.300 \text{ L}) = 0.150 \text{ mol}$.

    a. $\text{g NaOH} = 0.150 \text{ mol NaOH}\left(\frac{40.0 \text{ g}}{\text{mol}}\right) = 6.00 \text{ g}$

    b. $\text{mL HCl} = 0.150 \text{ mol HCl}\left(\frac{1000 \text{ mL}}{12.0 \text{ mol}}\right) = 12.5 \text{ mL}$

    c. $\text{L } NH_3 = 0.150 \text{ mol } NH_3\left(\frac{22.4 \text{ L}}{\text{mol}}\right) = 3.36 \text{ L}$

    d. $\text{mL } C_3H_6O = 0.150 \text{ mol } C_3H_6O\left(\frac{58.1 \text{ g}}{\text{mol}}\right)\left(\frac{1 \text{ mL}}{0.790 \text{ g}}\right) = 11.0 \text{ mL}$

33. It is frequently convenient in titration and other solution stoichiometry problems to use millimoles, mmol, rather than moles, since molarities are both mol/L and mmol/mL, and molecular masses are both g/mol and mg/mmol.

$$\text{mol HA} = (27.85 \text{ mL})(0.1094 \text{ mmol NaOH/mL})\left(\frac{1 \text{ mmol HA}}{1 \text{ mmol NaOH}}\right)$$

$$= 3.047 \text{ mmol} = 0.003047 \text{ mol}$$

$$M_{HA} = \frac{177 \text{ mg}}{3.047 \text{ mmol}} = 58.1 \text{ mg/mmol} = 58.1 \text{ g/mol}$$

35. The mass of nitrogen in the protein sample is readily obtained from

$$g_N = (0.2000 - 0.0742)\left(\frac{\text{mol HCl}}{L}\right)(0.5000 \text{ L})\left(\frac{1 \text{ mol } NH_3}{1 \text{ mol HCl}}\right)\left(\frac{1 \text{ mol N}}{1 \text{ mol } NH_3}\right)\left(\frac{14.007 \text{ g}}{\text{mol}}\right)$$

$$= 0.8810 \text{ g}$$

and the mass percent from

$$\% \, N = \left(\frac{0.8810\,g}{5.503\,g}\right) \times 100\% = 16.01\%$$

37. This exercise goes from a microscopic diameter, which might be obtained, for example, by scanning tunneling microscopy or X-ray diffraction, to the bulk substance. The density is

$$\rho = \frac{M}{N_A V_{\text{atom}}} \approx \frac{M}{N_A d^3} = \frac{28.086\,\text{amu}}{(6.022 \times 10^{23}\,\text{amu}/g)(2.72 \times 10^{-8}\,\text{cm})^3} = 2.32\,g/\text{cm}^3$$

The experimental density is 2.33 g/cm^3. To get the edge length of a 1-kg cube of Si, we note that if $V_{\text{atom}} \approx d^3$, then $V_{\text{cube}} = NV_{\text{atom}} = Nd^3 = nN_A d^3 = a^3$, where a is the edge length. Thus

$$a^3 = nN_A d^3 = 1000.\,g\left(\frac{1\,\text{mol}}{28.086\,g}\right)\left(\frac{6.022 \times 10^{23}}{1\,\text{mol}}\right)(2.72 \times 10^{-8}\,\text{cm})^3 = 431\,\text{cm}^3$$

$$a = 7.56\,\text{cm}$$

The same result could have been obtained from the computed density, $a^3 = m/\rho$, but this way the proportionality between atomic and macroscopic lengths is brought out.

39. As in the previous problem,

$$\bar{\varepsilon} = \frac{68.3\,\text{kcal}}{\text{mol}}\left(\frac{1000\,\text{cal}}{\text{kcal}}\right)\left(\frac{4.184\,J}{1\,\text{cal}}\right)\left(\frac{\text{mol}}{6.022 \times 10^{23}}\right) = 4.75 \times 10^{-19}\,J = 2.96\,\text{eV}$$

41. To accelerate the calculations we use the "engineering formula" developed in the text,

$$V(r) = \frac{kq_1 q_2}{r} = (14.40\,\text{eV}\,\mathring{A})\frac{q_1 q_2}{r}$$

where on the right the q_i are in units of e, the electronic charge. Thus

a. $V(\text{He}^{2+}\cdots e^-) = (14.40)\dfrac{(+2)(-1)}{0.31} = -93\,\text{eV}$

b. $V(p^+\cdots p^+) = (14.40)\dfrac{(+1)(+1)}{0.74} = +19.5\,\text{eV}$

c. $V(\alpha\cdots\text{Au}) = (14.40)\dfrac{(+2)(+79)}{1.00 \times 10^{-4}} = 2.28 \times 10^7\,\text{eV} = 2.28\,\text{MeV}$

d. $V(+2.52e\cdots e^-) = (14.40)\dfrac{(+2.52)(-1)}{0.84} = -43\,\text{eV}$

43. a. As the α-Au distance r goes to infinity, $E = K_\infty = 4.87$ MeV, while at the turning point r_0 for a head-on collision $E = V(r_0)$. Thus $V(r_0) = K_\infty$, and inverting the Coulomb potential yields

$$r_0 = k\frac{q_1 q_2}{V}$$

$$= 14.40\,\text{eV\,Å}\,\frac{(+2)(+79)}{4.87\times10^6\,\text{eV}}$$

$$= 4.67\times10^{-4}\text{Å}$$

b. See diagram on right. $r_{1/2}(K = K_\infty/2)$ may be readily read off the graph. To calculate $r_{1/2}$, we use $E = K + V$, and note that, since $E = K_\infty$, then $V(r_{1/2}) = K_\infty/2 = V(r_0)/2$. Then, since V and r are inversely proportional, $r_{1/2} = 2r_0 = 9.34\times10^{-4}$ Å

45. Each electron interacts with the other electron and the nucleus, giving

$$V_e = \frac{(-e)(-e)}{r_{ee}} + \frac{(+2e)(-e)}{r_{ne}} = 14.40\,\text{eV\,Å}\left(\frac{+1}{1\,\text{Å}} - \frac{2}{1\,\text{Å}}\right) = -14.40\,\text{eV}$$

We have again used the engineering Coulomb formula (Section 1.7). To get the total potential energy, we can't simply double the above, since that would count the electron-electron repulsion twice, but we can correct by subtracting one repulsion. Thus $V = 2V_e - e^2/r = -28.8 - 14.4 = -43.2$ eV. The potential energy can be decreased by moving the electrons apart; the best we can do is a collinear arrangement $e^-\text{----}He^{2+}\text{----}e^-$, yielding $V = -50.4$ eV. (Exercise 4.1 is related.)

2

The Quantum Revolution:
The Failure of Everyday Notions
to Apply to Atoms

Your Chapter 2 GOALS:

- Identify and describe electromagnetic radiation, commonly known as "light", using its two related attributes, frequency and wavelength
- Describe light and matter as having, simultaneously, wave-like and particle-like properties
- Outline the inability of classical physics to explain the radiant energy emitted by a "blackbody" and the photoelectric effect
- Understand the necessity of quantization in order to explain the behavior of light and matter on an atomic scale
- Use Bohr's ideas to relate line spectra to atomic structure
- Combine Bohr's and deBroglie's ideas about the atom to derive equations that give the energy and "radius" of electronic orbits of the hydrogen atom
- State the Schrödinger equation

Chapter 2 KEY EQUATIONS:

- 2.3, 2.6, 2.9, 2.16, 2.17
- ΔE(eV) = 1239.842 eV nm/λ(nm) $\approx$ 1240 eV nm/λ(nm) (See Exercises 3 and 12)
- $v = c/\lambda$
- $E = hv = hc/\lambda$

Overview

Who knew that history contained so much math! Well, science history does anyway. While we have limited ourselves to mostly algebra, which is adequate for an introduction to this material, know that the complete mathematical description of the quantum nature of matter is much more difficult than algebra and presents a stumbling block to many experienced chemists (and computers for that matter). We mention this to ask your persistence, not frighten you! In the interest of full disclosure, note that Chapter two contains the phrases: "Planck was uncomfortable with the quantum idea..." and "Schrödinger himself puzzled over the meaning of ψ for several years...and Albert Einstein rejected the whole business." That these scientific giants debated their hypotheses indicates, at least, two things. First, overturning the paradigm of Newton was very difficult. Second, it shows that they were truly scientists, never willing to accept fully anything even they themselves said. *Rational* skepticism is an asset for any scientist. The historical outline in Chapter 2, therefore, notes the key players and the times of their key assertions. What, exactly, they believed in their hearts and minds – and when – are much more difficult, if not impossible, facts to ascertain.

Exercises 1, 2, and 3 deal with the application of $v = c/\lambda$ and $E = hv$. Note that Maxwell could easily obtain the frequencies of Exercises 1 and 2 along with you. He knew that *any* electromagnetic wave obeys $v = c/\lambda$. The earth-shattering ability you have, that Maxwell did not, is to relate the frequency to the energy of the wave (or photon) by using $E = hv$. Exercise 3 can be performed quickly and filed away in your memory banks, but we urge you take the time to understand the derivation of the handy engineering formulas that are presented in this manual. These will save you time in the future and help you to learn what "normal" values are in chemistry. Whereas the value of 1240 is obtained in Exercise 3, a more precise value is derived for Exercise 12 by using the constants with all available significant figures. Exercises 6 through 8 are simple applications of the photoelectric effect equation, Equation 2.5, which can be written in the form $K = hv - W$. When answering Exercise 8, remember that photons are indivisible.

The work of brilliant scientists essentially reduces Exercises 9 through 14 to repeated use of the engineering formula derived in Exercise 3 or its more precise form of Exercise 12, $E(eV) = 1239.842$ eV nm / λ(nm). While these Exercises are computationally simple, their importance can't be overstated. Confined electrons (that is, electrons in atoms) exhibit wave-like properties. From Figures 2.9 and 2.11 this demands that the electrons exist as discrete standing waves, not having access to a continuum of energy levels. (These "standing waves" are already familiar to you as "orbitals.") The downward (upward) transitions between these energy levels produce (absorb) photons. Thereby, we have reduced 100 years of science, starting with Fraunhofer, to a few sentences. For Exercise 14(b), note that we often approximate the energy of an orbital as being equal to the negative of the ionization energy (IE) for that orbital. (See Section 4.5.) The signs are opposite because we consider a bound electron to have a negative energy (Equation 1.18), whereas the IE is viewed as the energy that we must *add* to an atom to cause ionization. Thus, IE > 0.

As you read Bohr's initial, planetary depiction of the atom, remember that he was wrong about many things. One reason, however, why his work is still revered is because of Equation 2.7. Today, Exercise 17, which verifies this equation, is done with ease. It might be followed by, "Wow. Look at that. Next Exercise please." In 1913, however, some of the constants (e, h, m_e) in this equation were becoming, but had not yet reached the point of "truth" with which we regard them today. Millikan's oil had dropped only a few years earlier (Figure 1.6), and h was still a mysterious entity. This connection, through R_H, further entrenched the "truth" of all of these physical constants. A second reason is that his derivation involved the quantization of the material world. Planck (1901) flirted with and Einstein (1905) clearly quantized *radiation*. Bohr, however, quantized the motion of the electron, which was viewed by all (and still is!) as *matter*. Quantizing its motion was a gigantic leap – one that would not be understood fully or accepted (even by Bohr himself) for more than a decade. Nonetheless, Bohr gave a *quantitative* explanation for line spectra that had puzzled science for a century, and he added his name to the growing list of scientists that were peddling the idea of quantized energy. Still, do not be confused. Bohr's circular orbits for H do NOT exist. The Bohr radius is a "characteristic length" in an atom. It (or multiples of it) is not a definite distance at which electrons reside in hydrogen or any other atom. Further, while Equation 2.17 gives the right energies for H, it is NOT based on the correct model of the hydrogen atom. That is not to say, though, that Equation 2.17 is not useful. It is very useful, conceptually and quantitatively. For example, see Equations 4.4 and 4.7. Also, with $Z = 1$, Equation 2.17 is used to solve Exercise 18!

The solutions to Exercises 19, 20, and 22 are related to Example 2.2. Note that as n_2 gets larger in Equation 2.3, we are describing progressively larger energy jumps. Eventually, the jump will be so large, the electron will have acquired so much energy, that it will be able to escape the hold of the nucleus. For Exercise 20 determine which transitions are within the visible spectrum, 1.8-3.1 eV, by writing an equation analogous to that of Exercise 15 via Equation 2.17. You should find that $\Delta E = 54.40$ eV$[(1/n_1)^2 - (1/n_2)^2]$. The needed equation for Exercise 22 is very similar to that of Exercise 20, only $Z = 42$.

Things really get weird when we move to the Matter Waves Exercises! Our experiences in life always lead us to believe that…well…matter is where it is. The mere concept that a piece of matter, no matter how small, can't be pinned down *exactly* is mind-boggling. (There is more to come on this in Section 3.5 with Heisenberg's Uncertainty Principle.) If you hoped to find the answer to the question 'What is an electron?' in this manual, we are sorry to report that no such answer is forthcoming. We do know that Schrödinger's equation gives us the correct energy for an electron, that electrons are negatively charged, that they are present in all atoms, that they jump around and emit photons, and so on. Is it a wave or a particle? We answer with a sincere

"YES!" How, then, are you to view the deBroglie wavelength of very small objects? It is best thought of as a measure of "quantumness." The smaller an object's deBroglie wavelength is, the less likely it is to exhibit observable quantum behavior. See Exercise 31 for "quantum" behavior which is 100% *not* observable. Be sure, however, to consider the dimensions of the object(s) with which the particle will interact. For example, the electron diffraction of Figure 2.10 is only seen if the electrons interact with "slits" or atoms that have spacing comparable to the deBroglie wavelength of the electron matter waves. Electrons are not the only matter that exhibit diffraction patterns. Two-slit experiments, somewhat like that shown in Figure 2.2, have shown that *individual* He atoms pass through both slits *at the same time* on their way to forming a diffraction pattern.

Exercises 23 through 28 require the application of Equation 2.9. Exercise 24 is a good example of how velocity changes can also alter the deBroglie wavelength. A typical velocity of a Na atom at 298 K is 52,000 cm s^{-1}. (See Equation 9.48.) The much smaller velocity in Exercise 24 results in a *really* cold group of atoms with deBroglie wavelengths that are much larger than the diameter of a "typical" sodium atom. At this velocity, then, the idea of a localized atom is meaningless. We see a similar result for Exercise 28, where the deBroglie wavelength becomes infinite at $18a_0$ since $E = V$ and $K = 0$ at this distance. Exercise 28 is difficult, but you should begin with the λ formula of Exercise 25. The only difficult step for Exercise 26 requires one to recognize that Equation 2.11 yields $\frac{1}{2}mv^2 = e^2/2r$ for use in $E = K + V$. The "inverse-square force field" of Exercise 27 refers to Equation 1.17 because $F \propto 1/r^2$.

Exercise 32 explores the difference between matter waves, for which $\lambda = h/[2mE_{particle}]^{1/2}$, and light waves, for which $\lambda = hc/E_{light}$.

1. In science, the response to the query "What's ν?" is always "c/λ." (Scientists have very low standards for puns.) Planck's hypothesis relates ν or λ to the photon energy E: $E = h\nu = hc/\lambda$. Thus

$$\nu = \frac{c}{\lambda} = \frac{2.998 \times 10^8 \text{ m/s}}{(1.932 \text{ Å})(10^{-10} \text{ m/Å})} = 1.552 \times 10^{18} \text{ s}^{-1} \text{ or } 1.552 \times 10^{18} \text{ Hz}$$

$$E = h\nu = (6.626 \times 10^{-34} \text{ Js})(1.552 \times 10^{18} \text{ s}^{-1}) = 1.028 \times 10^{-15} \text{ J} = 6417 \text{ eV}$$

The J → eV energy conversion should be familiar from Chapter 1. X-rays are of much shorter wavelength than visible light, and consequently their photons are of much higher energy.

3. When calculations are to be done for two or more values of an independent variable, all combinations of constants should be evaluated only once to save time and labor. Obtaining photon energies in eV from wavelengths is one of those situations.

$$E(\text{eV}) = \frac{hc}{\lambda} = \frac{(6.626 \times 10^{-34} \text{ Js})(2.998 \times 10^8 \text{ m/s})}{(1.602 \times 10^{-19} \text{ J/eV})(10^{-9} \text{ m/nm}) \lambda(\text{nm})}$$

$$= \frac{1240 \text{ eV nm}}{\lambda(\text{nm})}$$

This handy "engineering" formula can now be used repeatedly:

$$E_{400} = \frac{1240}{400} = 3.10 \text{ eV} \qquad E_{700} = \frac{1240}{700} = 1.77 \text{ eV}$$

If the answer is to be in J, all that changes is the "magic factor"; further formulas can be developed, but typically only a few of these formulas are frequently used. To obtain J, the J/eV conversion above is omitted, while J/mol or kJ/mol requires multiplying by Avogadro's Number N_A:

$$E(\text{J}) = \frac{hc}{\lambda} = \frac{(6.626 \times 10^{-34} \text{ Js})(2.998 \times 10^8 \text{ m/s})}{(10^{-9} \text{ m/nm}) \lambda(\text{nm})} = \frac{1.986 \times 10^{-16} \text{ J nm}}{\lambda(\text{nm})}$$

$$E(\text{kJ/mol}) = E(\text{J})(6.022 \times 10^{23} /\text{mol})(10^{-3} \text{ kJ/J}) = \frac{1.196 \times 10^5 \text{ (kJ/mol) nm}}{\lambda(\text{nm})}$$

$$E(\text{kcal/mol}) = \frac{E(\text{kJ/mol})}{4.184 \text{ kJ/kcal}} = \frac{2.859 \times 10^4 \text{ (kcal/mol) nm}}{\lambda(\text{nm})}$$

Substituting λ = 400 nm and 700 nm leads to

λ(nm)	E(J)	E(kJ/mol)	E(kcal/mol)
400	4.97×10^{-19}	299	71.5
700	2.84×10^{-19}	171	40.8

We shall find that these energies are of a "chemical" magnitude.

5. Solving Wien's Law for T,

$$T = \frac{0.288 \text{ cm K}}{\lambda} = \frac{0.288 \text{ cm K}}{(530 \text{ nm})(10^{-7} \text{ cm/nm})} = 5430 \text{ K}$$

In many problems it is convenient to characterize the light wave in terms of *wavenumbers*

(cm^{-1}) $\bar{\nu} = 1/\lambda$; here $\bar{\nu} = 18,900\ \mathrm{cm}^{-1}$. Wien's Law then reads $T = (0.288\ \mathrm{cm\ K})\bar{\nu}$.

7. Using Equation 2.5 and the results of Exercise 3,

$$K = \frac{hc}{\lambda} - W = \frac{1240\ \mathrm{eV\ nm}}{436\ \mathrm{nm}} - 2.1\ \mathrm{eV} = 2.84 - 2.1 = 0.7\ \mathrm{eV}$$

9. The two required energies are $\Delta E_{58.4} = 1240/58.4 = 21.2$ eV and $\Delta E_{2.06\mu m} = 1240/(2.06 \times 10^3) = 0.602$ eV (see Exercise 3). Since these lines originate from the same upper state, the locations of the lower states are determined relative to the original state by the computed ΔE's, as illustrated schematically on the right. As the diagram makes clear, $\Delta E_{12} = 21.2 - 0.6 = 20.6$ eV. (The energy levels of He involved have electron configurations $1s^2$, $1s^1 2s^1$, and $1s^1 2p^1$; see Chapter 4.)

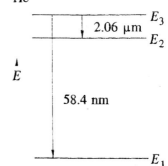

11. In eV, $\Delta E_{\mathrm{Li}} = 1240/671 = 1.85$ eV; similarly, $\Delta E_{\mathrm{Na}} = 2.10$ eV and $\Delta E_K = 1.62$ eV. ΔE in cm^{-1} is simply the inverse of the wavelength expressed in cm; since $10^7\ \mathrm{nm} = 1$ cm, $\Delta E(\mathrm{cm}^{-1}) = \bar{\nu} = 10^7/\lambda(\mathrm{nm})$. After you learn the approximate wavenumber range of visible light (14,000-25,000 cm^{-1}; see Figure 2.3), the "$1/x$" key on your calculator allows instant conversion of λ's to $\bar{\nu}$'s. Here $\Delta E_{\mathrm{Li}} = 10^7/671 = 14,900\ \mathrm{cm}^{-1}$, $\Delta E_{\mathrm{Na}} = 17,000\ \mathrm{cm}^{-1}$, and $\Delta E_K = 13,100\ \mathrm{cm}^{-1}$.

13. a. We use an improved version of the formula obtained in Exercise 3, and obtain the results

λ (nm)	610.441	612.391	616.387
ΔE (eV)	2.03106	2.02459	2.01147
$\Delta(\Delta E)$		0.00647	0.01312
		(ΔE_{21})	(ΔE_{32})

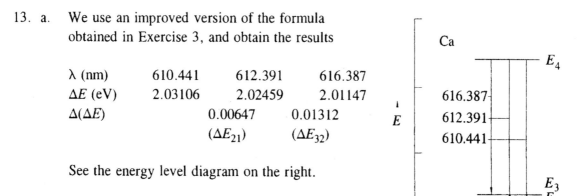

See the energy level diagram on the right.

b. The energy spacings between adjacent lower levels are given as $\Delta(\Delta E)$ above. We find

$1/\lambda_{21} = 52.2\ \mathrm{cm}^{-1}$, $\lambda_{21} = 192\ \mu m$, and
$1/\lambda_{32} = 105.86\ \mathrm{cm}^{-1}$, $\lambda_{32} = 94.5\ \mu m$,

both in the microwave region.

15. a. For the derivation, see Equation 2.18 and the discussion surrounding it. To cast the result in eV units, we note from the above solution to Exercise 3 that hc = 1240. eV nm = 1.240×10^{-4} eV cm, and therefore $hcR_\mathrm{H} = 13.60$ eV. We thus have

another "engineering formula":

$$\Delta E = \frac{hc}{\lambda} = 13.60 \, \text{eV}\left(\frac{1}{n_1^2} - \frac{1}{n_2^2}\right)$$

For Balmer-α, $n_1 = 2$ and $n_2 = 3$, yielding

$$\Delta E = 13.60 \, \text{eV}(\tfrac{1}{4} - \tfrac{1}{9}) = 1.889 \, \text{eV}$$

b.   The above formula implies that $E_n = -13.60 \, \text{eV} / n^2$; substituting $n = 2, 3,$ and 4 into this formula leads to the energies given at the left of Figure 2.12.

17. From Equation 2.7 and frontpage constants in cgs-esu,

$$R_\infty = \frac{2\pi^2 m_e e^4}{h^3 c} = \frac{2\pi^2(9.10938 \times 10^{-28} \, \text{g})(4.803204 \times 10^{-10} \, \text{esu})^4}{(6.62607 \times 10^{-27} \, \text{erg s})^3(2.99792458 \times 10^{10} \, \text{cm/s})}$$
$$= 109,737 \, \text{cm}^{-1},$$

where you should note that 1 $\text{esu}^2 \equiv 1$ erg cm. This constant is called R_∞ because it refers to an immobile, and therefore infinitely-massive, nucleus. To get R_H, we replace m_e by the reduced mass μ as given in Equation 2.8; since $\mu = 0.999456 m_e$, we may simply multiply the above result by 0.999456 to obtain $R_H = 109,677 \, \text{cm}^{-1}$, a value which agrees with the experimental one to six figures. You may also use mks units in this calculation, provided you replace e^4 in Equation 2.7 by $k^2 e^4$, where $k = 8.98755 \times 10^9$ J m C^{-2} is the Coulomb's Law constant. (To be fair, it must be noted that the modern values of m_e, e, and h are partially determined by R_H—one of many input data—making the present calculation somewhat redundant, whereas Bohr's original estimate was independent.)

19. For $n_2 \to \infty$, we replace n_1 by a general n, and Equation 2.3 reduces to

$$\frac{1}{\lambda} = R_H\left(\frac{1}{n^2} - 0\right) \quad \text{or} \quad \lambda = \frac{n^2}{R_H} = \frac{h^3 c n^2}{2\pi^2 m_e e^4}$$

where we have used Bohr's result for the Rydberg constant, Equation 2.7. In absorption, an atom will pass from state n to a state where the electron may depart to infinite distance (see the previous Exercise), that is, the atom ionizes, H $\to$ H$^+$ + e$^-$. The reverse process is called "radiative recombination", and is believed to be the principal mechanism by which ions neutralize themselves in interstellar space.

21. The potential energy in H is $V = -e^2/r$, while in He$^+$ it is $V = -2e^2/r$. From Exercises 15 and 20, the energies are both -13.60 eV. Thus we have the energy diagram shown on the next page, which for H is identical to the "bound" case shown in Figure 1.10. From the diagram (or the Bohr derivation itself), $V = 2E$. Since the Es are equal, the Vs must also be equal, or $e^2/r_H = 2e^2/r_{He^+}$. Thus $r_{He^+} = 2r_H$ is required to yield the same potential energy and total energy. Note that the kinetic energy $K = -E$ for both systems. The relations $V = 2E$ and $K = -E$ are each consistent with the Virial Theorem (see Exercise 27).

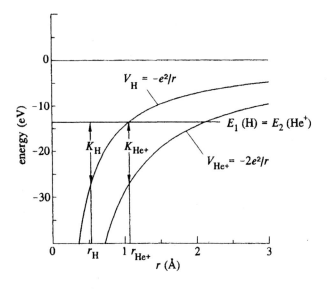

23. In Chapter 1 we learned of "thermal energy"; Equation 1.15 allows us to estimate the average thermal energy of the electron at temperature T by $\bar{\varepsilon} \approx k_B T$. We may further assume that this is kinetic energy, $\bar{\varepsilon} = K$, and use the formula developed in Example 2.6 to write

$$\lambda = \frac{h}{mv} = \frac{h}{\sqrt{2mK}} = \frac{h}{\sqrt{2mk_B T}}$$

$$= \frac{6.626 \times 10^{-34}\,\text{J s}}{\sqrt{2(9.109 \times 10^{-31}\,\text{kg})(1.381 \times 10^{-23}\,\text{J/K})(1500\,\text{K})}} = 3.4 \times 10^{-9}\,\text{m} \;(34\text{Å})$$

Here it is worth noting that $k_B T$ at 1500 K is 2×10^{-20} J or about 0.1 eV.

25. If $E = K + V(r)$ then $K = E - V(r)$ (see Exercise 1.42); substituting this into the result of Example 2.6 and Exercise 23 yields at once

$$\lambda(r) = \frac{h}{\sqrt{2m[E - V(r)]}}$$

This relationship was exploited by Schrödinger in developing the wave equation; see Exercise 30.

27. Using the formula of Exercise 25 and the Virial Theorem in the form $2E = V$, we can write

$$\lambda = \frac{h}{\sqrt{2m(E-V)}} = \frac{h}{\sqrt{-2mE}} = h\left[\frac{n^2 h^2}{4\pi^2 m^2 e^4}\right]^{1/2} = \frac{nh^2}{2\pi m e^2}$$

Equation 2.13 was used in the third step. This agrees with the result of Exercise 26.

29. Taking $mvr = nh/2\pi$ and multiplying by $2\pi/mv$ leads to $2\pi r = nh/mv = n\lambda$, which is Equation 2.10.

31. We find $n = 2\pi r/\lambda = 2\pi(1.5\times10^{13}\text{cm})/(3.7\times10^{-61}\text{cm}) = 2.5\times10^{74}$. This number exceeds the *cube* of N_A, but is not as big as a googol (10^{100}). A $\Delta n = 1$ transition is not detectable, yielding a change in the radius of earth orbit of $\lambda/2\pi = 10^{-61}$ cm.

3

Wave Mechanics and the Hydrogen Atom: Quantum Numbers, Energy Levels, and Orbitals

Your Chapter 3 GOALS:

- Recognize that the Schrödinger equation contains a very important type of operator, the Hamiltonian operator
- Identify the two parts of the Hamiltonian operator as really being two operators, namely the kinetic and potential energy operators for a system
- Rationalize the conversion from Cartesian coordinates to spherical polar coordinates in facilitating the solution of the Schrödinger equation
- Identify the two functions that comprise ψ, the radial (R) function and the angular (Y) function
- Define the three quantum number, n, l, and m, and use them to construct wave functions for the hydrogen atom from Table 3.1
- Be able to draw sketches of radial functions, R_{nl}, boundary surface diagrams of orbitals, and RDF's
- Calculate the radial and angular nodes of wave functions by examining the radial, r, and angular, θ and ϕ, dependence
- View RDF's as a much more realistic picture of electron location in an atom than Bohr's orbits
- Write the Uncertainty Principle in two different forms and use it to approximate atomic properties
- Be able to explain why electrons don't collapse into the nucleus as Maxwell's theory predicts

Chapter 3 KEY EQUATIONS:

- 3.1, 3.6, 3.14, 3.15
- 3.13, conceptually
- $RDF = r^2[R(r)]^2$
- The ability to form linear combinations as in Equation 3.10

Overview

Indeed, the weird behavior of matter waves continues in this chapter. To avoid mental anguish, we must find *something* on which we can "hang our hats", something that is definite, unlike the electron's location, simultaneous knowledge of the electron's position and momentum, the m values for orbitals that have been combined via a linear combination, ad infinitum. The best hope for this comes from the boundary conditions of Section 3.1. Mercifully, quantum mechanics can't prevent the electron from being somewhat rational. The electron must be somewhere, and it must be near the nucleus. (Isn't it amusing that quantum mechanics has us celebrating such simple things?) These facts, however, allow the Schrödinger equation to be solved exactly for hydrogen. We do not obtain the unsavory result for the energy that we do for position. *We know the energy of the hydrogen atom exactly for every value of n.* Well, so did Bohr! True enough, but Bohr's description explained very little (from a modern viewpoint). Quantum mechanics *explains* (with the extension of the H model) what Mendeleyev and Moseley found with experimental brilliance. It *explains* why the electron is quantized without the mere *supposition* that the angular momentum is quantized, why electrons are diffracted by slits or atoms as in Figures 2.10 and 3.10, and why selection rules exist (see Section 3.7 and Exercise 28). Thus, we can revel in the certainty that quantum mechanics correctly yields the energy of the hydrogen atom and explains (reasonably well) chemical behavior in all its forms: the order of the periodic table, bonding, spectroscopy, etc. What we can't do is say what an electron is. We must live with this dual being and describe it as best we can.

It is important to highlight the difference between the probability density $|\psi|^2$ and the RDF. From Figure 3.3a the thoughtful student will surely notice that ψ_{100}, and therefore $|\psi_{100}|^2$, (or any s orbital) is a maximum at $r = 0$, yet the 1s RDF is zero there (Figure 3.9). Is this a contradiction? Indeed, it is not because any $|\psi|^2$ is proportional to the probability of finding an electron *at a point in space*, whereas the RDF is proportional to the probability of finding the electron *anywhere in a very thin, spherical shell of thickness dr*. Therefore, these two quantities convey fundamentally different information. For example, in Exercise 18 maximizing (or plotting) the 1s RDF tells us that the *most probable radius* is a_0, yet visual inspection of Figure 3.3(a) tells us that the *most likely point* is at the origin, or nucleus. Likewise, we use the square of the wave function (a la Born) for Exercise 17 and the RDF for Exercise 20 to show that the most probable points in a 2p orbital are at $\pm 2a_0$ and the most probable radius is at $4a_0$. These values ($\pm 2a_0$ and $4a_0$) are simultaneously true, giving different pieces of information about the many places an electron visits in its travels.

Now, since RDF's give the most probable radius for an electron, RDF's neatly fit our semi-planetary/shell structure model of the atom and are often invoked for showing "what's going on" with the electron. Note from Figure 3.9 that, as n grows, the RDF's grow in amplitude at distances farther from the nucleus, implying that *on average* a 3s orbital is farther from the nucleus than a 2s orbital, *on average* a 4p orbital is farther from the nucleus than a 3p orbital, etc. Since we can make these distance calculations, we can still conceptually think about the energy of the orbitals in terms of Equation 1.18.

As a classical particle approaches a turning point (see Example 1.4 at the point r_0), its kinetic energy approaches zero. If it were to somehow continue past this point, its kinetic energy would become negative. Exercise 19 deals with the quantum effects that allow electrons to tunnel into these regions where classical particles can't go. Now, since tunneling is, perhaps, the epitome of quantum behavior, we expect that Newtonian mechanics will be utterly incapable of explaining it. A predicted negative (and impossible) kinetic energy is, therefore, not an unexpected consequence of Newton's ignorance of the wave properties of massive particles.

Those students who have good backgrounds in chemistry are, perhaps, confused that we say a 2s and 2p orbital are degenerate. How can this coincide with *every* electron configuration that I have ever written, you might ask. As you answer Exercise 3, keep at the forefront of your thoughts that Chapter 3 discusses the hydrogen atom only. *Only for hydrogen is the quantum number n the sole arbiter of energy ordering.* When the hydrogen model is extended to other atoms (see Figure 4.7), the familiar ordering emerges, and we must look at l as well as n. For this reason Exercise 4 uses the same formula that is given in Figure 3.2 for a 4p or 100s state. Note that there is no l dependence, only n.

To help with the sketch of the orbital described in Exercise 9, note that the angular nodes are found by setting the angular portion of the wave function equal to zero,

$$\left(\frac{5}{3}\cos^3\theta - \cos\theta\right) = 0,$$

and solving for θ. Likewise for $Y_{20}(\theta)$ of Exercise 12. Similarly, Exercise 11 requires finding the nodes for $R_{30}(r)$ through use of the quadratic equation on the factor $(27 - 18\rho + 2\rho^2)$. The equations necessary for Exercise 10 are in the summary.

To obtain the exact values in the text answer key for Exercise 22, include all of the factors. Thus, use $\Delta p_x = h/4\pi\Delta x$ and $K \approx (\Delta p_x)^2/2m$. This Exercise illustrates that confinement produces a significant amount of zero-point kinetic energy, K_{zp}, only when $K_{zp} \approx k_B T$, which only occurs when m is very small, via the K equation just given. The watermelon "particle's" large mass prohibits it from exhibiting quantum behavior and, consequently, from having $K_{zp} \approx k_B T$. For a Be atom, the significant K_{zp} is energy that can't be shared with other Be atoms. It is energy that the Be atom possesses simply because of its confinement, and this energy leads to uncertainty in its momentum and position. Exercise 23 continues the theme of wave-particle duality, indicating that "wave packets", or localized particles, spread out as they travel, an effect called dispersion. We obtain such an outlandish answer because we allow the electron to travel a large distance. Since electrons travel much shorter distances in their daily lives, such a large spatial breadth is never seen. Note that for Exercise 24 an accurate graph should yield a larger uncertainty than the approximate a_0 used for the text answer key. Further, this Exercise shows that the famous 13.6 eV energy for the $1s$ orbital is, in fact, a zero-point energy.

Exercise 28 deals with selection rules, and you may exclude the possibility of transitions within the $n = 4$ level. Exercises 29 and 30 are done in the same manner. We only add that the spread in transition wavelength is derived from the wavenumber relation as follows

$$\bar{v} = \frac{1}{\lambda} \quad \Rightarrow \quad \frac{d\bar{v}}{d\lambda} = -\frac{1}{\lambda^2} \quad \Rightarrow \quad d\bar{v} = -\frac{1}{\lambda^2}d\lambda,$$

followed by the change from d's to Δ's.

3 Wave Mechanics and the Hydrogen Atom

1. See the chapter summary for the allowed values of n, l, and m, and the text answer key.

3. For the hydrogen atom, the energy depends only on the n value. If orbitals have the same n value, they have the same energy. Thus, 200 and 210 are degenerate, and 322 and $31-1$ are degenerate. (When magnetic effects are included, these states show small energy splittings, that is, they are no longer precisely degenerate.)

5. The work w must raise the energy of the atom from $E_1 = -13.6$ eV to $E = 0$; since work is being done on the system, $w = -E_1 = +13.6$ eV $=$ IE. By the Virial Theorem (Exercise 2.27), $V = 2E$, and thus V must be increased from -27.2 eV (-2IE) to 0, corresponding to complete separation of the proton and electron $(r \to \infty)$.

7. Using Equation 3.10 as a pattern, we expect that

$$3p_x = \frac{1}{\sqrt{2}}(3p_{+1} + 3p_{-1})$$

$$3p_y = \frac{1}{i\sqrt{2}}(3p_{+1} - 3p_{-1})$$

Taking the wave functions from Table 3.1 and using Euler's formula, we get

$$3p_x = \frac{1}{\sqrt{2}}(R_{31}Y_{11} + R_{31}Y_{1-1})$$

$$= \frac{R_{31}}{\sqrt{2}}\left(\frac{\sqrt{3}}{2\sqrt{2\pi}}\right)\sin\theta(e^{i\phi} + e^{-i\phi})$$

$$= R_{31}\frac{\sqrt{3}}{2\sqrt{\pi}}\sin\theta\cos\phi$$

$$= \frac{4}{81\sqrt{6}}a_0^{-3/2}(6-\rho)\rho e^{-\rho/3}\frac{\sqrt{3}}{2\sqrt{\pi}}\sin\theta\cos\phi$$

$$= \frac{\sqrt{2}}{81\sqrt{\pi}}a_0^{-3/2}(6-\rho)\rho e^{-\rho/3}\sin\theta\cos\phi$$

where the Ys are inserted in line 2, Euler's formula applied in line 3, and R_{31} inserted in line 4. An analogous procedure may be followed for $3p_y$. Note that at $\theta = 0$ or $\phi = \frac{\pi}{2}$ (the y-z plane), the orbital has zero amplitude; and that the combination $r\sin\theta\cos\phi = x$ appears as a factor.

9. Using the rule $\psi_{nlm}(r,\theta,\phi) = R_{nl}(r)Y_{lm}(\theta,\phi)$, we find

$$\psi_{430} = \left(\frac{1}{768\sqrt{35}}a_0^{-3/2}\rho^3 e^{-\rho/4}\right)\left(\frac{3\sqrt{7}}{4\sqrt{\pi}}\left(\frac{5}{3}\cos^3\theta - \cos\theta\right)\right)$$

$$= \frac{1}{1024\sqrt{5\pi}}a_0^{-3/2}\rho^3 e^{-\rho/4}\left(\frac{5}{3}\cos^3\theta - \cos\theta\right)$$

Based on the labels $2p_z$ and $3d_{z^2}$, $4f_{z^3}$ is suggested. See Figure 17.17 for a boundary surface; this also points along z, but displays two "spare tires," that is, a nodal plane (x-y) and two nodal cones (at $\theta = 39.2°$ and $140.8°$).

11. For making hand sketches (highly recommended for this exercise) it is helpful first to find the nodal positions for each $R(r)$ by finding the roots of $R(r) = 0$ in units of a_0, and marking the r-axis off in the same units. Then, with some attention to the form of the functions shown in Figures 3.3, 3.4, and 3.5, and with computed or estimated values at extrema used to mark off the R-axis, reasonably accurate sketches can be made. Alternatively you can use your favorite graphics or spreadsheet software to compute and plot the functions using the entries in Table 3.1. See the "official" graph at the right, in which the vertical scale was chosen to emphasize the large-r domain. Note that, as in Figure 3.4, $-R_{20}$ has been plotted, to make the long-range tail positive. The graphs help to convey nodal structure and the dependence of orbital size on n.

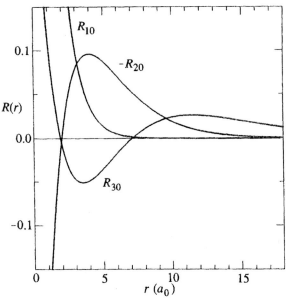

13. According to Table 3.2 the $3p_z$ orbital function is

$$3p_z = \frac{\sqrt{2}}{81\sqrt{\pi}} a_0^{-3/2} (6-\rho)\rho e^{-\rho/3} \cos\theta$$

Nodes occur when a factor becomes zero. The premultiplying normalization constants are never zero, nor does the exponential ever become zero except as $r \to \infty$. The radial factor shows zeroes at $\rho = 0$, and at $6 - \rho = 0$ or $r = 6a_0$; the latter is a spherical node. The angular factor is zero when $\cos\theta = 0$ or $\theta = \frac{\pi}{2}$, the x-y plane; this is the angular node, and it gives the orbital its shape.

15. Refer to the orbital figures. You will not find $4s$ or $4p_z$ shown there, but their shapes are determined by the Y functions, and they will be larger than the orbitals of smaller n. Again, you need not include the spherical nodes in $4p_z$, though you should be able to say how many such nodes there are.

17. This exercise is interpreted as requesting point(s) of maximum electron density. Since the orbital functions ψ are products of independent functions of r and of angles, maxima in ψ^2 may be found by setting (partial) derivatives of ψ^2 with respect to r and angles to zero separately. It is also useful to note that, e.g., $d(\psi^2)/dr = 2\psi d\psi/dr$, and that it thus suffices to set $d\psi/dr$ to zero to find extrema; as noted in the exercise, premultiplying constants may be ignored when solving $d\psi/dr = 0$ for r. For the radial part we use the reduced distance ρ. For $2p_z$, using Table 3.2 we then have

where A is a premultiplying constant. To find the angles of the maxima, we first note that $2p_z$ does not depend on ϕ, and that therefore the extrema must also be independent

$$2p_z = \psi_{210} = \frac{1}{4\sqrt{2\pi}} a_0^{-3/2} \rho e^{-\rho/2} \cos\theta$$

$$\frac{d\psi}{d\rho} = A \frac{d}{d\rho}\left(\rho e^{-\rho/2}\right) = 0$$

$$A(1 - \rho/2) e^{-\rho/2} = 0$$

$$\rho = 2 \quad or \quad r = 2a_0 \doteq 1.06\,\text{Å}$$

of ϕ. For θ we have

$$\frac{d\psi}{d\theta} = B \frac{d}{d\theta}(\cos\theta) = 0$$

$$B(-\sin\theta) = 0$$

$$\theta = 0, \pi$$

Thus there are two maxima, at $(r,\theta,\phi) = (2a_0,0,\phi)$ and $(2a_0,\pi,\phi)$ for any ϕ in $(0,2\pi)$. Except perhaps for the distance at the extrema, they could have been deduced from the form of the $2p_z$ boundary surface.

19. In Exercise 18 a graph was requested for the $1s$ radial distribution function, reproduced here, for which a block-counting method was suggested for estimation the area under this curve. The area beyond the classical turning point may be estimated using the same block counting method to be $0.05(2+1+1) = 0.2$; the exact value based on evaluation of the

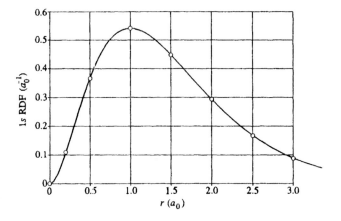

integral is 0.238. The integral can be written in terms of ρ, and evaluated by repeated integration by parts. You should also learn to make effective use of integral tables for your future work in science; the integral you need here will be found there in generalized form. To quote Owl, "The customary procedure in such cases is as follows."

$$P(r \geq 2a_0) = \int_{2a_0}^{\infty} r^2 [R_{10}]^2 dr = 4 \int_{2}^{\infty} \rho^2 e^{-2\rho} d\rho$$

From tables

$$\int x^m e^{ax} dx = \frac{m!\, e^{ax}}{a^{m+1}} \sum_{n=0}^{m} (-1)^{m-n} \frac{(ax)^n}{n!}$$

Here $m=2$ and $a=-2$, giving

$$P = 4 \frac{2!\, e^{-2\rho}}{(-2)^3}\left(1 - (-2\rho) + \frac{(-2\rho)^2}{2!}\right) \Bigg|_{2}^{\infty}$$

The upper limit ∞ yields zero due to the exponential; thus

$$P = e^{-4}(1 + 4 + 8) \doteq 0.238$$

21. The RDFs for $l = n-1$ are of the form RDF $\propto \rho^{2n} e^{-2\rho/n}$. Differentiating yields

$$\frac{d(\text{RDF})}{d\rho} = \left(2n - \frac{2\rho}{n}\right)\rho^{2n-1}e^{-2\rho/n} = 0$$

and $\rho = n^2$ follows at once. The Bohr theory yielded $r = n^2 a_0$ [Equation 2.16], identical to the present result. As discussed in Section 3.6, the corresponding classical orbits are closest to the circular Bohr orbits for these cases.

23. Referring to Figure 3.10(a), passing the electron through a hole of diameter Δx gives rise to a transverse velocity component $v_x \approx h/(m\Delta x)$. In the time $t = d/v_z$ that it takes the electron to travel to the moon a distance d away, it will also have traveled a distance $x = v_x t$ "off course"; x is roughly the size of the "spot" the electron makes on the moon. We thus have

$$x = v_x t \approx \frac{h}{m\Delta x}\frac{d}{v_z} = \frac{hd}{m\Delta x}\left[\frac{m}{2K}\right]^{\frac{1}{2}} = \frac{d}{\Delta x}\frac{h}{\sqrt{2mK}}$$

$$x = \frac{(2.3 \times 10^5 \text{ mi})(1609 \text{ m/mi})}{1 \times 10^{-3} \text{ m}} \frac{6.6 \times 10^{-34} \text{ kg m}^2/\text{s}}{[2(9.1 \times 10^{-31} \text{ kg})(10 \text{ eV})(1.6 \times 10^{-19} \text{ J/eV})]^{\frac{1}{2}}}$$

$$= 143\text{m} \approx 100\text{m}$$

Here it is also of interest to ask for intermediate results, such as travel time t (200 s), travel velocity v_z (2×10^6 m/s), and transverse velocity v_x (0.7 m/s). What we have computed is actually the spatial breadth of the "wave packet" representing the traveling electron after travel time t.

25. Since $l = 1$ for any p orbital, we can compute K_{rot}:

$$K_{\text{rot}} = \frac{1(2)(6.626 \times 10^{-34} \text{ J s})^2}{8\pi^2(9.109 \times 10^{-31} \text{ kg})[(4)(0.5292 \times 10^{-10} \text{ m})]^2(1.602 \times 10^{-19} \text{ J/eV})}$$

$$= 1.70 \text{ eV}$$

Then we use the Virial Theorem $K = -E = -(-3.40 \text{ eV}) = +3.40 \text{ eV}$ to find $K_{\text{vib}} = K - K_{\text{rot}} = 3.40 - 1.70 = 1.70 \text{ eV}$ as well. Taking K_{vib} to represent the zero-point motion in the radial coordinate, we employ the Uncertainty Principle governing the radial coordinate to find

$$\Delta r = \frac{h}{4\pi\Delta p_r} \approx \frac{h}{4\pi\sqrt{2mK_{\text{vib}}}}$$

$$\Delta r = \frac{6.626 \times 10^{-34} \text{ kg m}^2/\text{s}}{4\pi\sqrt{2(9.109 \times 10^{-31} \text{ kg})(1.70 \text{ eV})(1.602 \times 10^{-19} \text{ J/eV})}}$$

$$= 7.49 \times 10^{-11} \text{ m} \approx 1.4a_0$$

Δr is thus of the same order as the radius of the Bohr orbit; it is expected to represent the width of the RDF for the $2p$ state.

27. First consider the classical frequency ν_n:

$$\nu_n = \frac{\upsilon_n}{2\pi r_n} = \left(\frac{2\pi e^2}{nh}\right)\left(\frac{4\pi^2 me^2}{2\pi n^2 h^2}\right) = \frac{4\pi^2 me^4}{h^3 n^3}$$

where Equation 2.16 and the results of Exercise 2.26 have been used. Now the transition frequency frequency $\nu_{n,n-1}$:

$$\nu_{n,n-1} = \frac{\Delta E_{n,n-1}}{h} = \frac{2\pi^2 me^4}{h^3}\left(\frac{1}{(n-1)^2} - \frac{1}{n^2}\right) = \frac{2\pi^2 me^4}{h^3}\left(\frac{2n-1}{n^2(n-1)^2}\right)$$

This formula is not quite the same as that for ν_n, but if we let n grow large, the "-1"s in numerator and denominator of the parenthetical factor become negligible, reducing it to $(2/n^3)$, and making ν_n and $\nu_{n,n-1}$ identical.

29. From Equation 3.15 we have

$$\Delta E \geq \frac{h}{2\pi\tau} = \frac{6.626 \times 10^{-34}\,\text{J s}}{2\pi(2.13 \times 10^{-9}\,\text{s})} = 4.95 \times 10^{-26}\,\text{J} = 3.09 \times 10^{-7}\,\text{eV} = 0.00249\,\text{cm}^{-1}$$

The spread in transition wavelength arises because $\bar{\nu} = 1/\lambda$ and, by taking differentials, $\Delta\bar{\nu} \approx -\Delta\lambda/\lambda^2$ or

$$\Delta\lambda \approx \lambda^2|\Delta\bar{\nu}| = (121.6\,\text{nm})^2(0.00249\,\text{cm}^{-1})(10^{-7}\,\text{cm/nm}) = 3.68 \times 10^{-6}\,\text{nm}$$
$$\Delta\lambda/\lambda = 3.03 \times 10^{-8}$$

4

Atoms with Many Electrons
and the Periodic Table

Your Chapter 4 GOALS:

- Be able to identify the term in Equation 4.1 that arises from electron-electron repulsion and understand that this term makes the Schrödinger equation impossible to solve *exactly* for *any* atom except hydrogen
- Know that *all* orbitals other than those of hydrogen are approximations, assumed to be nearly identical to those found for the single electron in hydrogen
- Define screening and use it to rationalize the energy ordering of orbitals in atoms with Z > 2
- Identify spin as the fourth quantum number and describe how spin can be used to explain doublets in atomic spectra (Figures 2.4, 2.8, and 4.2), the results of the Stern-Gerlach experiment (Figure 4.3), and doubly occupied orbitals
- Use the Pauli Principle, the Aufbau Principle, and Hund's Rule to construct orbital occupation diagrams
- Relate orbital occupation diagrams to the *s*-, *p*-, *d*-, and *f*-block structure of the Periodic Table
- Obtain electron configurations directly from the Periodic Table
- Define atomic radius, ionization energy, electron affinity and describe the trends they exhibit in the Periodic Table
- Describe the use of iteration in computer modeling

Chapter 4 KEY EQUATIONS:

- 4.1, 4.7
- 4.8, conceptually

Overview

Parts of Chapter 4 give us a respite from the intensive mathematical treatment of the first three chapters. Additionally, you might be thinking, 'Ah. Finally. Some chemistry.' Hopefully, though, we have shown that the first three chapters are also chemistry. Before the respite, however, let's discuss the math. Again.

It is likely that your introduction to atomic structure in previous courses contained no reference to a wave function. This puts you in a class with nearly *every* other student who has taken chemistry, your authors included. Wave functions are, however, the basis of our modern description of matter, and discussion of them must occur if you are to see the whole picture. In your salad days, when you were green in judgment (see Shakespeare, *Antony and Cleopatra*, actually), the orbitals in which you placed electrons were envisioned as empty shells, perhaps, or no thought at all was given to *what* an orbital is. The fact that 2 electrons go into the '1*s* orbital of He', as it is often (incorrectly) phrased, is simply accepted as a truth. The secrecy is more than understandable, given the six-coordinate, mathematical train wreck that is Ψ_{He}. In the introduction to Chapter 3 in this manual, we said that quantum mechanics explained the order of the Periodic Table. We say this because confined waves, which are described by quantum mechanics, are what we call orbitals. By assuming that all electrons in atoms occupy similar orbitals (1*s*, 2*s*, 2*p*,...) with the same general shapes and nodal structure and that each electron has an intrinsic spin, all of the atoms that react readily with water line up (Group IA), all of the nonmetal atoms that typically form − 1 ions line up (Group VIIA), all of the atoms that essentially do not react line up (Group VIIIA), etc. You already knew these groups lined up. Further, it is likely that you would attribute this to the identical valence electron configuration in each group.

(A more than valid answer, by the way.) *Now you should have a much deeper understanding that these electron configurations consist of products of orbitals that are the result of the wave properties of electrons (or are the waves themselves) and that a mathematical device called a many electron wave function is used to obtain as much information as we can from these orbitals.* While you might not yet understand all of the math or the philosophical implications (few, if any, do), your list of "descriptors" for atomic behavior has grown in a fundamental way.

The extension of the one-electron hydrogen orbitals to other atoms is explored in Exercises 1 through 5 and 20 through 26. Exercise 1 is a difficult problem, incorporating many of the concepts of the first four chapters. It extends Exercise 26, Chapter 2 and revisits the virial theorem of Exercise 27, Chapter 2. The remaining Exercises in this group quantify the effects of screening through Z_{eff}, which is the end result of electron-electron repulsion and nuclear attraction. Z_{eff} is the positive charge that the electron under consideration actually feels. Exercise 4 is particularly important, as it extends the analysis shown in Figure 4.4 for the $n = 2$ orbitals to the $n = 3$ orbitals. For a given n the lower the l value for an orbital, the more time it spends near the nucleus, thereby lowering its energy due to Coulombic attraction. s orbitals penetrate through the core electrons best, p orbitals next, etc.. Consequently, s orbitals exist at the lowest energy for a given n, followed by p, d, and f orbitals. Thus, the familiar energy level ordering used to construct electron configurations emerges. From Figure 4.7 note that this holds within each n level without exception. The "mix-ups" in the order come between orbitals of different n (Figure 4.8).

For Exercise 20 Z_{eff} is obtained from just above Equation 4.3 in the textbook. The given equation for $E(He)$ is just Equation 4.4, and since $E(He^+)$ is given by the Bohr formula, use Equation 2.17 with $Z = 2$ and $n = 1$ to calculate it. Note that the radius of He, as calculated in part (b), is smaller than that of H, as expected from the larger Z_{eff}. Exercise 26 is a stark example of the need for approximations when solving the Schrödinger equation. For U Equation 4.8 would contain 4186 electron-electron repulsion terms. Now, if *one* of these terms makes the He Schrödinger equation unsolvable, then imagine what 4186 terms do! Exercises 21 through 25 form a series of Exercises that incorporate Z_{eff}, IE, and electron configuration, serving as a semi-review of Chapter 4 topics. Take a moment to look at the coefficients in the Z_{eff} equation of Exercise 21. You should quickly notice that they increase as the electrons they multiply lie further within the core of the electron cloud, implying that electrons in lower n levels do a better job of screening than the electrons within the same n as the electron under consideration. The electrons in the $n - 2$ shell, are assumed to screen (block, if you prefer) a full unit of nuclear charge, the same estimate that is made in Exercise 3. The electrons in the $n - 1$ shell only screen 0.86 of a nuclear charge. Electrons in the same n level only screen 0.35 of a nuclear charge. This compares well with the 0.31 value used in Exercise 2. Thus, Slater's Z_{eff} equation qualitatively matches what we expect based on the smaller radial extent of core electrons. Note, however, the warning in Exercise 23 that tells us Slater's rules don't consider electrons with the same n but different l. In fact, we are given the rules for s and p electrons only. That l is important to screening leads us to the "chemistry" of Chapter 4 and a discussion of the lanthanide contraction.

Equation 4.4 indicates that orbital energy is determined by the competition between Z_{eff} and n. Generally, the increase in n is of greater importance in determining orbital energy than the increase in Z_{eff} as one descends a group in the Periodic Table, resulting in the general trend in ionization energy that is shown in Figure 4.9. As we move from the $4d$ to $5d$ metals, however, the Z_{eff} increase is of greater importance than the n increase. As a specific example, let's compare Hf and Zr. Observe from Figure 4.6 how long Hf must "wait" for its "last" electron as the $4f$ orbitals fill. These 14 added $4f$ electrons do not screen the increasing Z very well, however, because of the small amplitude of their RDF's in the core of the atom. Thus, the outer electrons of some of the 5d metals are held more tightly than the outer electrons of the 4d metals, yielding the "lanthanide contraction", and the radii of the 5d metals are much less than expected. Remarkably, Hf is *smaller* than Zr! For this same reason, the IE of Hf is greater than that of Zr. This exception to the general *IE* trend is seen for the elements from $Z = 72$ to $Z = 80$. (We didn't consider the confusing comparison of Y to La or Lu because La is anomalous and both La and Lu have a 5d^1 configuration.)

Exercise 16 calculates "Mulliken" electronegativities, which differ from the more familiar Pauling electronegativities of Table 5.1. This points out that the atomic properties that we discuss are absolutely dependent on our definitions of them. For Exercise 18 beware that Cu is anomalous, and for Exercise 7 beware that U is anomalous.

4 Atoms with Many Electrons and the Periodic Table

1. In this exercise you are asked to treat each of the electrons in "choreographed He" separately. This makes the problem similar to the Bohr-deBroglie atom of Chapter 2; the difference lies in the electron-electron repulsion, which provides an additional term in the Coulomb force and potential energy. Using the fact that $r_{12} = 2r$, Newton's Second Law (Equation 2.11) becomes

$$-\frac{2e^2}{r^2} + \frac{e^2}{(2r)^2} = m\left(-\frac{v^2}{r}\right) \quad or \quad r = \frac{7}{4}\frac{e^2}{mv^2}$$

Equations 2.9 and 2.10 still combine to yield $nh/(mv) = 2\pi r$; substituting this new result for r and solving for v yields

$$v_n = \frac{7\pi e^2}{2nh} = \frac{7}{4}\frac{2\pi e^2}{nh}$$

$$r_n = \frac{n^2 h^2}{7\pi^2 m e^2} = \frac{4}{7}n^2 a_0$$

$$\lambda_n = \frac{8\pi}{7} n a_0 = \frac{4}{7} 2\pi n a_0$$

Thus v is greater, r smaller, and λ shorter than in H, as expected from the greater nuclear attraction. These changes, however, are not as great as in He$^+$—where the factor 7/4 is replaced by 2—owing to our approximate account of e$^-$–e$^-$ repulsion. The total energy according to Equation 4.1, considered as a classical energy, is

$$E = \frac{1}{2}mv_1^2 + \frac{1}{2}mv_2^2 - \frac{2e^2}{r_1} - \frac{2e^2}{r_2} + \frac{e^2}{r_{12}}$$

$$= mv^2 - \frac{7}{2}\frac{e^2}{r}$$

where the second line follows from setting $v_1 = v_2$, $r_1 = r_2 = r$, and $r_{12} = 2r$, and combining terms. Now Newton's Law above says $mv^2 = \frac{7}{4}e^2/r$; this makes

$$E = \frac{7}{4}\frac{e^2}{r} - \frac{7}{2}\frac{e^2}{r} = -\frac{7}{4}\frac{e^2}{r}$$

$$E_n = -\frac{7}{4}e^2\left(\frac{7}{4n^2 a_0}\right) = -\frac{49}{8}\frac{e^2}{2a_0 n^2}$$

where in the second line we inserted the expression for r_n above. Comparing this result with twice Equation 4.4 (one term for each electron), we have a hydrogen-like model with $Z_{eff} = 7/4 = 1.75$; this compares well with the "best" Z_{eff}, 1.69, quoted in the text. Using $e^2/(2a_0) = 13.6$ eV, we find $E_n = -83.3$ eV/n^2, yielding a ground-state energy $E_1 = -83.3$ eV, to be compared with the experimental result -79.0 eV, an error of only 5%. From the last set of relations above it is seen that $V = -2K = 2E$, showing the validity of the Virial Theorem for Coulomb systems with more than two particles.

3. For the 2p electron in He $1s^1 2p^1$, we suppose that $Z_{eff} = 2 - 1 = 1$. Then we may apply Equations 4.7 to compute r_{2p} and E_{2p}:

$$r_{2p} = \frac{2^2 a_0}{1} = 4a_0 = 2.117 \,\text{Å} \quad \text{and}$$

$$E_{2p} = -\frac{(1)(14.40 \,\text{eV Å})}{2(2.117 \,\text{Å})} = -3.401 \,\text{eV}$$

These results are identical to those for an H atom in the $2p$ state, and also close to the experimental value for E.

5. Neglecting electron-electron repulsion in He is equivalent to allowing each electron to see the full, unscreened nuclear charge, $Z = 2$. The energy of He is then twice the energy of each electron, identical to that of two He^+ ions. The energy of He^+ is readily computed using the Bohr formula Equation 2.17, and is also given in the text to be -54.4 eV; thus the energy of He without e^-—e^- repulsion is $2(-54.4) = -108.8$ eV. This is in serious disagreement with experiment, -79.0 eV, implying that neglecting e^-—e^- repulsion, or, equivalently, screening, is not feasible.

7. One of the species (U) is anomalous; the text answer key gives only the predicted configuration. See Table 4.1 for U.

9. The first step is to choose the ground state; only (b) obeys both Pauli and Hund. Secondly, only (c) violates Pauli, by putting electrons of identical spin into the $3s$ and $3p_1$ orbitals, and is therefore impossible. The rest must be excited states by elimination. The specification of these excited states may be considered in more detail in a later course.

11. No energy splitting between l-sublevels implies hydrogenic degeneracy of all levels of a given n. For H, the orbital degeneracy is n^2, and hence the electron capacity is $2n^2$. For $n = 5$, the capacity is 50 electrons, and hence 50 distinct elements will occur in the fifth period. The noble gas for $n = 5$ would have completely filled l subshells for $l = 0$-4, and hence the valence configuration $...5s^2 5p^6 5d^{10} 5f^{14} 5g^{18}$. The earlier shells $n = 1$-4 would also be filled, accounting for the first $2+8+18+32 = 50$ elements. Thus the $n = 5$ noble gas would be element #100, a transuranic element not occurring naturally. It is also speculated that the chemistry of the elements would be more highly variable, with electrons in the valence shell more easily able to take on s, p, d, ... character, or a combination of these ("hybrid orbitals"; see Chapter 6) as their chemical environment demands. This would eliminate the l-block structure of the periodic table.

13. The IE trend, increasing from lower left to upper right, may be thought of as a staircase leading roughly from Cs to He, along an upward-sloping diagonal. The steps of this staircase must extend perpendicular to the direction of ascent, roughly along downward-sloping diagonals; along each step we expect similar IEs. The only exceptions to this should come from elements with especially stable electron configurations, as discussed in the text. Based on this analysis, we expect elements in the downward-sloping diagonal beginning with C: C, P, Se, I, to have similar IEs. Se $(...4p^4)$ is somewhat low, being one electron beyond the half-filled p subshell; this electron is forced to pair with another,

raising the total energy.

15. An atom A* with an anomalously large IE has an especially stable electron configuration (filled or half-filled shell or subshell). An electron cannot be added to such a configuration without destroying the special stability, and hence the process $A^{*-} \to A^* + e^-$ will require less energy than if A were not especially stable. The energy change in this process is defined to be the EA; hence the EA will be anomalously low.

17. For the atomic radius, group anomalies do not occur until the sixth period (lanthanide contraction) and period anomalies occur only in the latter half of the transition series. Thus the general trend (radius decreases from lower left to upper right) suffices for the elements chosen:

   a. $(Ne) < F < S < Ga < Ba$ b. $(Ar) < Cl < Br < I < Te$ c. $N < Si < Al < Mg < Cs$

   There are no known covalent bonds involving Ne or Ar; their placement in the series is consistent with the trend in radius.

19. Ca^{2+} and Sc^{3+} both have [Ar] closed-shell configurations, making them isoelectronic with P^{3-}, S^{2-}, Cl^-, and K^+. Here you are encouraged to reason as Mendeleev did in interpolating/extrapolating properties for undiscovered elements. We refer to Figure 4.13. For Ca^{2+}, we observe that the ionic radii for the ions of Groups I and II increase roughly linearly going down the group, with a step size of ~ 0.3 Å, while going across a period they decrease in ~ 0.2 Å steps. The former gives an estimate of 0.95 Å, while the latter yields 0.97 Å; averaging these gives our estimate $r(Ca^{2+}) = 0.96$ Å. The experimental value from crystallographic analysis is 0.99 ± 0.03 Å. For Sc^{3+}, the same trends hold, and we get (less certain) estimates of 0.80 Å looking down Group III, and 0.77 Å looking across Period 4; the average yields $r(Sc^{3+}) = 0.79$ Å. The experimental value is 0.73 ± 0.03 Å.

21. For Ar, $N_n = 8$, $N_{n-1} = 8$, and $N_{n-2} = 2$, yielding

   a. $Z_{eff}(Ar) = 18 - [0.35(8-1) + 0.86(8) + 1.00(2)] = 6.67$
      $r_{3p} = n^2 a_0 / Z_{eff} = 3^2(0.5292 \text{ Å}) / 6.67 = 0.714$ Å

   b. $Z_{eff}(Ar^+) = Z_{eff}(Ar) + 0.35 = 7.02$, yielding
   $$\frac{IE}{E_H} = -(8-1)\frac{(7.02)^2}{3^2} - \left(-(8)\frac{(6.67)^2}{3^2}\right) = 1.2165$$

   and IE $= 1.2165(13.60 \text{ eV}) = 16.5$ eV. The tabulated value is 15.8 eV, for an error of $+4\%$.

23. For second-row neutral atoms A beyond Li, $N_n = Z-2$, $N_{n-1} = 2$, and $N_{n-2} = 0$, while for the corresponding cations $Z_{eff}(A^+) = Z_{eff}(A) + 0.35$. Thus, for C we have N and O are analogous. Comparing with tabulated values,

$$Z_{eff} = 6 - [0.35(4-1) + 0.86(2)] = 3.23$$

$$\frac{IE}{E_H} = -(4-1)\frac{(3.58)^2}{2^2} - \left(-(4)\frac{(3.23)^2}{2^2}\right) = 0.8206$$

$$IE = (13.60\,eV)(0.8206) = 11.2\,eV$$

$$r = 2^2\frac{a_0}{3.23} = 0.66\,\text{Å}$$

atom	IE_{calc}	IE_{expt}	r_{calc}	r_{expt}
C	11.2	11.26	0.66	0.77
N	12.6	14.53	0.55	0.70
O	13.8	13.62	0.47	0.66

Agreement for IE is good except for N, where Slater's rules miss the extra stability of the half-filled shell. The computed radii are all smaller, suggesting that the bond lengths reflected in r_{expt} do not allow for interpenetration (overlap) of orbitals up to their most probable radii, given by r_{calc}. Further, r_{calc} decreases more rapidly than r_{expt} in the C,N,O sequence, suggesting that bonding is less intimate for the later atoms in the period.

25. For Na 3s, $Z_{eff} = 11 - [0.35(0) + 0.86(8) + 1.00(2)] = 2.12$; $r = 9a_0/2.12 = 2.25$ Å
    For Na$^+$ 2p, $Z_{eff} = 11 - [0.35(7) + 0.86(2)] = 6.83$; $r = 4a_0/6.83 = 0.31$ Å

The dramatic difference between both Z_{eff}'s and r's reflects the enormous gap between n-shells of an atom in both radius and energy. The radii are much further apart than those in Fig 4.13 (1.86 and 0.85 Å, respectively), due to the difference in bond types used to fix the experimental radii. Here $r(Na^+)$ from experiment is much larger than the orbital radius computed here owing to its origin in ionic bond lengths. When IE(Na) is computed from the Slater formula of Exercise 21, the single valence electron causes the ionic term to drop out, and IE $= -E(3s) = 13.60$ eV $(2.12)^2/9 = 6.8$ eV, 32% larger than the experimental 5.14 eV of Figure 4.9. Again, the absolute error is < 1.7 eV, comparable to that in He. Inspecting the results of the last 5 exercises, the absolute errors range from <0.1 eV (C) to 1.9 eV (N).

5

Valence Electron Configurations, Periodicity, and Chemical Behavior

Your Chapter 5 GOALS:

- Identify the three regions of the Periodic Table: metals, nonmetals, metalloids
- Justify the noble gas rule using electron configurations and Equation 5.6
- Describe the key difference between the arbitrary concepts of ionic and covalent bonding as the degree of electron sharing between the atoms involved in the bond
- Be able to draw Lewis structures and calculate bond orders from them
- View actual bonding as a spectrum of sorts, relying on electronegativity differences
- Write mechanisms for simple reactions
- Classify reactions as acid-base or oxidation-reduction and into smaller sub-classes
- Become familiar with many reaction types
- Describe the three definitions of acid-base reactions: Arrhenius, Brønsted-Lowry, and Lewis
- Balance complicated oxidation-reduction reactions by the process listed after Equation 5.52
- Define an oxidation number and assign it to various elements or compounds
- Predict the solubility of a compound based on the rules of Section 5.7 and use this knowledge to write ionic equations

Chapter 5 KEY EQUATIONS:

- 5.6, 5.8, 5.31, 5.35, 5.38, 5.39, 5.46
- It's a semi-equation! The reaction classification scheme of Section 5.7

Overview

You will notice that for a chapter with so many equations (61, officially), not many are in the "Key Equations" list. This is simply due to many of the equations being examples of specific reaction types or mechanisms. The ones that we listed are essentially fundamental definitions of overarching significance. While Chapter 4 contained some of the familiar "chemistry" that you know, Chapter 5 represents the Mother lode (not to be confused with the Comstock lode, Exercise 51) of familiar chemistry.

Exercises 1, 2, and 3 require the noble gas rule and the assumption that metal/nonmetal reactions lead to ionic compounds. Exercise 4 can be attacked by calculating the two putative molar masses for the unknown metal X that result for the two possible outcomes of the reaction, XCl and XCl_2. Exercises 6 and 7 explore the second Chapter 5 goal via Equations 5.5 and 5.6. The electron configuration part should be very easy, but the (probably) new aspect for you is that we ask you to quantify the ionic bond. A lone Na atom is in no hurry to lose its $3s$ electron, evidenced by its ionization energy. Since the ionization requires energy *input*, there must be something that yields an energy *release*. This "something" is the attainment of noble gas configuration *and the Coulomb attraction*, which is determined by Equation 1.18.

Exercises 9 through 14 ask you to draw Lewis structures, and rules for drawing them are given in Example 5.2 and the Exercises themselves. Some students might be troubled by differences between the structures that we present in *University Chemistry* and those that have been seen in other introductory courses. First, we treat all valence electrons as equals in the compound and do *not* use different symbols (for example, ×'s or •'s) to keep track of which electrons come from which atom. The electrons certainly lose track of where they originate, and the Lewis structure should reflect that. Second, we repeat that it is our practice not to violate the octet rule unless we absolutely must (Exercise 11). Thus, we draw SO_2 as shown before Example 5.2 and do not form a double bond between the P atom and the non-hydroxyl O atom in H_3PO_4, as some chemists do. In such cases of competing Lewis structures some authors utilize the concept of "formal charge" to decide which structure is best, and you will undoubtedly encounter this term if you remain in chemistry. We feel, however, that adherence to the octet rule is sufficient and better approximates the "true" bonding that is present. Exercises 15, 16, and 17 rely on the arbitrary designations of ionic, polar covalent, and pure covalent that are given in the *Electronegativity* subsection of Section 5.2 and Equations 5.7 and 5.8.

Predicting reaction products can be very difficult because it requires one to identify what types of reactants are involved and how they usually react. Exercises 18, 24, 29, 33, and 40 demand that you review the various reaction types that appear throughout Chapter 5. For example, for Exercise 18(d), compare Equation 5.16. For Exercise 24(g), note that Section 5.4 states that SiO_2 is inert to attack by water or acid. Each part of these Exercises requires you to go on a scavenger hunt of sorts for the exact information or *analogous* information. An example of an analogous reaction is that needed for Exercises 24(f) and (i). Equations 5.22 and 5.23 are quite analogous to these Exercises and provide material for an educated guess as to the products of the reactions. These are just a few examples of the effort required to answer Exercises 18, 24, 29, 33, and 40 and the accompanying mechanism and classification Exercises. While this type of learning is necessary, it is not the only way to become a good prognosticator. You can also make a list of general facts, not specific reactions that can guide your predictions. Such a list might *begin* with the following five facts: i. most Group 1 and 2 metals give away electrons in redox reactions with water; ii. metals and nonmetals form ionic bonds with each other; iii. nonmetals share electrons with each other in covalent bonds; iv. metals react with water to give basic solutions and $H_2(g)$; v. nonmetals react with water to give acidic solutions. Fact iv., for example, gives you 8 (maybe 9) reactions! Your intuition and ability to presage reaction products increase with the number of facts that you know. Learning a list of facts, therefore, is hardly pointless.

In Exercise 34 you are asked to write ionic equations (some chemists use the term net ionic equations). This process is important in that it reveals the real actors in a reaction, shedding light on the actual chemistry at work, and is a necessary tool when it comes to balancing redox reactions. While this topic was explored in Equations 5.48 and 5.49, it was not dwelled on since this is a procedure that should have been mastered in a previous course. Note that solids, such as $NaHCO_3(s)$ and $Na_2S(s)$, and liquids, such as $H_2O(l)$, are not "broken up". As usual, though, chemistry throws the frequent curve ball. Exercise 34(f) has $Ca(OH)_2$ listed as an aqueous reagent, even though the solubility rules indicate otherwise. To explain this, we point out that "soluble" (usually) means 1 g or more will dissolve in 100 mL of water. The designation "insoluble" doesn't mean that *none* of the compound goes into solution. Keep this in mind as you solve Exercise 53 and when we get to Section 12.4. This same Exercise 34(f) also contains a weak acid, $H_3PO_4(aq)$. Often these are not broken up, but the approximation used to not break the acid up gets worse as the acid gets "less weak". To break up or not to break up, then, is at times a matter of chemical perception. Note that this reaction is classified as an Arrhenius acid-base reaction, and we can achieve an overall ionic equation of $H^+(aq) + OH^-(aq) \rightarrow H_2O(l)$, as expected for an Arrhenius acid-base reaction, if $H_3PO_4(aq)$ is broken up.

Exercises 26, 31, and 32 all explore the role of electronegativity χ in chemical behavior. Exercises 26 and 31 examine acidic or basic behavior via the tug-of-war over electrons in various bonding arrangements due to a $\Delta\chi$ that exists between the tug-of-war participants, with the more electronegative element winning the war. This procedure is satisfactory and leads to logical and *correct* conclusions. When the $\Delta\chi$ strategy is attempted for Exercise 32, however, the exactly wrong conclusion is reached. Indeed, relying solely on $\Delta\chi$ to predict acidic and basic properties is to travel the primrose path of chemistry. Exercise 32 suggests that bond strength, *not* $\Delta\chi$, is the controlling factor in determining the acid strength of the hydrogen halides in aqueous solution. This is yet another example of the care that must be exercised when accepting the explanations that are given for chemical phenomena. Without thought, one can hear the explanations for Exercises 31 and 32 and accept them as "true" and correct. It is the nagging thought of 'Why does $\Delta\chi$ work for Exercise 31 but bond energy is needed for Exercise 32?' that is at the foundation of chemistry. Chemists and all physical scientists seek that *one* correct explanation. The explanation that can "cover all the bases". For now we must accept different explanations for seemingly very similar events. (Our experience with the electron, however, has prepared us for this.)

Classifying acid-base reactions is no easy matter. As a first example, let's look at Exercise 38 for which there are no mixtures of simple aqueous acids and bases. Consequently, the Arrhenius classification will not be used. The reactions for which a lone pair on an oxygen attacks a hydrogen atom on water, as in Equation 5.20, are classified as Brønsted-Lowry acid-base reactions. Reactions that mirror Equations 5.24 and 5.25, such as Exercise 24(b), are classified as Lewis acid-base reactions. What this reveals is that classifying reactions often requires knowledge of the *mechanism* of the reaction. With experience this becomes easier, but, if you are stuck, write out the mechanism for the reaction before you try to classify it. As a second example let's look at Exercises 55 and 56. Within seconds you can easily identify every redox reaction in these two Exercises due to the consumption or formation of an element. With that you are well on your way, but, as with Exercise 38, the acid-base reactions are harder to classify. Exercise 55(h) is the most difficult, but $Al_2(SO_4)_3(aq)$ solutions are, in fact, acidic. If the reaction is not redox, however, you can deduce this by the process of elimination and the fact that you know that $OH^-(aq)$ is a base. Further, the Lewis classification can be used if one views each hydroxide as donating an electron pair to Al in the $Al(OH)_4^-(aq)$ product. This type of product is called a complex and is explored further in Section 12.3 and Chapter 17.

The balancing of combustion reactions (Exercise 41) follows the "extended inspection" process of Section 5.6, but it might involve first using a factor of ½ in front of the O_2, followed by multiplying the entire equation by 2. See also Equation 18.2. For Exercise 41(h) follow the C, H, O rule with N balanced last. The balancing of redox reactions in Exercises 45, 46, 48, 49, 50, and 51 follows the steps outlined in Section 5.6. While this may seem like a large number of steps, you will likely have the steps memorized after completing a few of these Exercises. By far, the hardest step is finding the right couples to use from the ionic equation. Assigning oxidation numbers and looking for changes in oxidation state is a typical way to achieve this. Solving all of the parts of Exercise 50 with the half reaction method is good practice for Chapter 13. For Exercise 50(d) the couples are $NH_4^+ \rightarrow N_2$ and $NO_2^- \rightarrow N_2$, as it is an inverse disproportionation reaction. Exercise 50(g) is also inverse disproportionation. For Exercise 50(e), the hardest one, the couples to use are $ClO_3^- \rightarrow Cl^-$ and $H_2O \rightarrow O_2$, with the second being fictitious. For Exercise 50(j) the couples are $SiF_6^{2-} \rightarrow Si$ and $K \rightarrow K^+$.

5 Valence Electron Configurations, Periodicity, and Chemical Behavior

1. Here we take advantage of the generality of the noble gas rule.

    a. $Rb([Kr]5s^1) \rightarrow Rb^+([Kr]) + e^-$ oxidation
    $Br([Ar]4s^23d^{10}4p^5) + e^- \rightarrow Br^-([Kr])$ reduction
    In the case of singly-charged ions, these half-reactions can simply be added. RbBr is the product from the reaction $Rb + Br \rightarrow RbBr$, ignoring for the time being the fact that Br is normally found as Br_2, a diatomic molecule.

    b. $Al([Ne]3s^23p^1) \rightarrow Al^{3+}([Ne]) + 3e^-$ oxidation
    $S([Ne]3s^23p^4) + 2e^- \rightarrow S^{2-}([Ar])$ reduction
    Here the oxidation half-reaction must be multiplied by 2, and the reduction by 3, in order to give the electrically neutral product Al_2S_3, and a reaction $2Al + 3S \rightarrow Al_2S_3$, when added.

    c. $Li([He]2s^1) \rightarrow Li^+([He]) + e^-$ oxidation
    $N([He]2s^22p^3) + 3e^- \rightarrow N^{3-}([Ne])$ reduction
    Multiply the oxidation by 3 and add to give the reaction $3Li + N \rightarrow Li_3N$; as in part (a), we ignore the normal form of N, N_2.

    d. $Ca([Ar]4s^2) \rightarrow Ca^{2+}([Ar]) + 2e^-$ oxidation
    $O([He]2s^22p^4) + 2e^- \rightarrow O^{2-}([Ne])$ reduction
    No multipliers are needed to give $Ca + O \rightarrow CaO$, ignoring O as O_2.

    e. $Sr([Kr]5s^2) \rightarrow Sr^{2+}([Kr]) + 2e^-$ oxidation
    $F([He]2s^22p^5) + e^- \rightarrow F^-([Ne])$ reduction
    Multiply reduction by 2 and add to give $Sr + 2F \rightarrow SrF_2$, ignoring F as F_2.

    This procedure for analyzing redox reactions will be generalized in later exercises.

3. Based on the noble-gas rule, $Y([Kr]5s^24d^1)$ should be in the form Y^{3+}, $Ba([Xe]6s^2)$ should be Ba^{2+}, and $O(1s^22s^22p^4)$ should be O^{2-}. Since the $YBa_2Cu_3O_7$ compound is overall neutral, the sum of the charges on these ions, $3 + 2(2) + 7(-2) = -7$, must be balanced by a +7 charge on the 3 Cu ions, or +2.33 per Cu. In an ionic compound, we expect the ionic charges to be integers; the arrangement of charges that minimizes the largest one is $2Cu^{2+} + Cu^{3+}$. Compounds of this type are the only examples where Cu^{3+} appears to exist.

5. We need Equation 1.18 in its engineering form, and Figure 4.13, for "touching" ions; in addition the conversion factors $1 eV = 23.0606 kcal/mol = 96.485 kJ/mol$ are useful here.

    a. $Li^+ \cdots F^-$ $V = \dfrac{e(-e)}{r_e} = -\dfrac{e^2}{r_+ + r_-} = -\dfrac{14.40 \, eV \, Å}{(0.53 + 0.98)Å} = -9.54 \, eV$
    $(-220 \, kcal/mol, -920 \, kJ/mol)$

b. $Cs^+ \cdots I^-$ $V = \dfrac{e(-e)}{r_e} = -\dfrac{14.40\,eV\,\text{\AA}}{(1.41 + 1.88)\text{\AA}} = -4.38\,eV$

$$(-100.9\ \text{kcal/mol},\ -422\ \text{kJ/mol})$$

In this case we estimated the Cs^+ ionic radius by linear extrapolation of the Group I cation radii. See Exercise 19 of Chapter 4.

c. $Na^+ \cdots S^{2-}$ $V = \dfrac{e(-2e)}{r_e} = -\dfrac{2(14.40\,eV\,\text{\AA})}{(0.85 + 1.84)\text{\AA}} = -10.7\,eV$

$$(-247\ \text{kcal/mol},\ -1030\ \text{kJ/mol})$$

d. $Ba^{2+} \cdots O^{2-}$ $V = \dfrac{(2e)(-2e)}{r_e} = -\dfrac{4(14.40\,eV\,\text{\AA})}{(1.23 + 1.40)\text{\AA}} = -21.9\,eV$

$$(-505\ \text{kcal/mol},\ -2110\ \text{kJ/mol})$$

Using the Cs^+ radius estimated in (b) and subtracting 0.20Å yields $r(Ba^{2+}) =$ 1.21Å. The value used here is based on crystalline properties.

e. $F^- \cdots Ca^{2+} \cdots F^-$

$$V = 2\frac{(2e)(-e)}{r_+ + r_-} + \frac{(-e)(-e)}{r_- + 2r_+ + r_-}$$

$$= -\frac{7(14.40\,eV\,\text{\AA})}{2(0.98 + 0.99)\text{\AA}} = -25.6\,eV$$

$$(-590\ \text{kcal/mol},\ -2470\ \text{kJ/mol})$$

Here we have 3 Coulomb terms: 2 attractions and 1 repulsion between the end anions. The arrangement of charges is identical to that in the "choreographed He" atom of Exercise 1, Chapter 4. See Exercise 21, Chapter 4 for $r(Ca^{2+})$.

7. We use the estimate of $r(Cs^+)$ from Exercise 5.(b) above, $r(Cl^-)$ from Figure 4.13, and IE(Cs) and EA(Cl) from Figure 5.6:

$$r_x = \frac{14.40\,eV\,\text{\AA}}{(3.89 - 3.62)\,eV} = 53.3\,\text{\AA}$$

$$D_e = \frac{14.40\,eV\,\text{\AA}}{(1.41 + 1.50)\,\text{\AA}} - 0.27\,eV = 4.68\,eV \quad (107.9\ \text{kcal/mol})$$

In this case the ionic model comes much closer to experiment; this is due mainly to the increased influence of higher-order terms in the attraction relative to NaCl or LiF, which cancels part of the error arising from neglecting the repulsion.

9.-13. Rules for drawing Lewis structures are given in the text and in each Exercise. See the text answer key for structures.

15. To determine the polarity of an A—B bond, we calculate $\Delta\chi = \chi(B) - \chi(A)$; by convention, $\chi(B) \ge \chi(A)$, and B therefore carries the partial negative charge $-\delta$. Using data from Figure 5.6, we find:

Bond	Sb–Br	B—O	C—S	Ca—P	N—H	Al—I	Rb—F	Si—Te	P—H	Ga—As
$\Delta\chi$	0.9	1.5	0.1	1.2	0.8	1.1	3.2	0.2	0.0	0.4
sense	+ −	+ −	+ −	+ −	− +	+ −	+ −	+ −	...	+ −

In order of increasing polarity, we have

Covalent: P–H C–S Si–Te *Polar covalent:* Ga—As N–H Sb–Br Al–I Ca–P B–O *Ionic:* Rb—F

where the polar-covalent criterion $0.4 \leq \Delta\chi \leq 1.7$ has been applied.

17. We use Equations 5.7 and 5.8 to find:

$$\Delta = D(HF) - [D(H_2)D(F_2)]^{1/2} = 135 - [(103)(37)]^{1/2} = 73.3$$

$$\chi(F) - \chi(H) = 0.21\Delta^{1/2} = 0.21(73.3)^{1/2} = 1.80$$

Taking $\chi(F) = 4.00$, $\chi(H) = 4.00 - 1.80 = 2.20$.

19. Those that yield H^+ are acidic, OH^- basic. See text answer key.

21. Which Al electron is chosen matters little for mechanistic purposes; the process must be repeated twice more to yield Al^{3+}.

23. Based on electronegativities, the order is:

$$Cs \sim Rb > Sr \sim Ca > Al > Ga$$

Al > Ga is anomalous, since the general trend for χ is to increase from lower left to upper right of the periodic table. The anomalously low $\chi(Al)$ has its origin in the IE anomaly associated with the $3s^2 3p^1$ configuration, one electron beyond a closed subshell. Ga is also in IIIA, but it occurs after $3d$ filling, which increases its Z_{eff} relative to Al, and gives it a higher-than-expected IE and χ.

25. See text answer key.

27. When Lewis formulated his dot structures, he had no knowledge of s/p subshell structure, and hence could not explain why the elements listed show more than one oxide. In each case O is the more electronegative element, and if it therefore always acquires 2 extra electrons in these oxides, the compound stoichiometry then depends on whether the central atom donates all valence electrons, or only the more labile p electrons. This changes the noble gas Ng rule to an "alkaline earth Ae rule" in the p-only compounds. For C to become like Be, it must lose 2 (not 4) electrons, leading to CO (instead of CO_2). Based on this scheme, we form the table:

Element	Valence config	Formula of oxide Ng rule	Ae rule	
C	$2s^2 2p^2$	CO_2	CO	
P	$3s^2 3p^3$	P_2O_5	P_2O_3	
N	$2s^2 2p^3$	N_2O_5	N_2O_3	(also N_2O, NO, NO_2, NO_3)

S	$3s^2 3p^4$	SO_3	SO_2
As	$4s^2 3d^{10} 4p^3$	As_2O_5	As_2O_3
Pb	$6s^2 4f^{14} 5d^{10} 6p^2$	PbO_2	PbO
Se	$4s^2 3d^{10} 4p^4$	SeO_3	SeO_2

See the text answer key for Exercise 12 for NO_x Lewis structures; for N_2O, we put a N in the center, and derive $:N\equiv N—\overset{..}{\underset{..}{O}}:$.

29. See text answer key.

31. See Equation 5.33. The lone pairs of water are shown there to attack the partially positive end of the H—F bond. In the present case, the lone pairs of water would have to attack the H atom in the H—S bond or the H atom in the H—Cl bond to produce H_3O^+. Since $\chi(Cl) > \chi(s)$, the H—Cl bond is more polar than H—S. The greater polarity of HCl makes the H atom more attractive to nucleophiles, and also enhances heterolytic cleavage leading to ionization.

33.-35. See text answer key.

37. See text answer key for labels; mechanisms are given at the right.

c.

i.

j.

39. Before reaction with the antacid
    $n_{HCl} = (20.00 \text{ mL})(0.5000 \text{ mmol/mL})$
    $= 10.00 \text{ mmol.}$
    After reaction, the NaOH titration of the remaining HCl (1:1 titration stoichiometry) shows
    $n_{HCl} = (26.43 \text{ mL})(0.2000 \text{ mmol/mL})$
    $= 5.286 \text{ mmol remaining.}$
    The difference is the moles consumed by the antacid,
    $$n_{HCl} = 10.00 - 5.286 = 4.71 \text{ mmol, giving}$$
    $$g_{HCl} = 4.71 \text{ mmol } (36.46 \text{ mg/mmol}) = 172 \text{ mg} = 0.172 \text{ g.}$$
    Thus, $g_{HCl} / g_{antacid} = 0.172 \text{ g} / 0.3960 \text{ g} = 0.434 \text{ g} / \text{g}$

41.,43. See text answer key.

45. A critical step in balancing aqueous redox reactions by the half-reaction method is the construction of an appropriate unbalanced ionic equation as a starting point. Only aqueous species (*aq*) can be written in ionic form, and common anions such as NO_3^- or SO_4^{2-} should not be broken down further. See Table 5.2 if you are unsure. $O^{2-}(aq)$ does not exist. Some compounds such as $H_2S(aq)$ are weak electrolytes, and should not be written in ionic form, although if you do, you will still obtain a properly balanced equation in the end after spectator ions have been reintroduced. Bear in mind that not all

spectator ions, nor all necessary acid, base, or water, may appear in the reaction in unbalanced form.

a. There are no spectator ions here initially. The ionic equation is:

$$CuS(s) + H^+(aq) + NO_3^-(aq) \rightarrow Cu^{2+}(aq) + SO_4^{2-}(aq) + NO(g)$$

Examination indicates that only S and N are undergoing a change in oxidation state. The half-reaction couples are therefore:

$$CuS \rightarrow Cu^{2+} + SO_4^{2-} \quad \text{and} \quad NO_3^- \rightarrow NO$$

The H^+ is omitted in stating the couples, since any H^+ or H_2O required will appear in the course of balancing the half reactions according to the stepwise procedure given in Section 5.6. The H^+ could have been omitted from the starting equation above as well. For the $CuS/CuSO_4$ couple (added species in boldface):

step 2: $CuS + \mathbf{4H_2O} \rightarrow Cu^{2+} + SO_4^{2-}$
step 3: $CuS + 4H_2O \rightarrow Cu^{2+} + SO_4^{2-} + \mathbf{8H^+}$
step 5: $CuS + 4H_2O \rightarrow Cu^{2+} + SO_4^{2-} + 8H^+ + \mathbf{8e^-}$.

Thus we find this is an oxidation (something that could also have been deduced from oxidation numbers). For the NO_3^-/NO couple:

step 2: $NO_3^- \rightarrow NO + \mathbf{2H_2O}$
step 3: $NO_3^- + \mathbf{4H^+} \rightarrow NO + 2H_2O$
step 5: $NO_3^- + 4H^+ + \mathbf{3e^-} \rightarrow NO + 2H_2O$,

a reduction. Note that steps (a) and (d) are not needed here. To balance electrons gained and lost, we multiply the oxidation by 3 and the reduction by 8 before adding, obtaining

$$3CuS + 12H_2O + 8NO_3^- + 32H^+ \rightarrow$$
$$3Cu^{2+} + 3SO_4^{2-} + 24H^+ + 8NO + 16H_2O .$$

Note that H^+ and H_2O come from both half-reactions, and thus appear on both sides of this ionic equation. Cancelling the excess H^+ on the left and H_2O on the right leaves

$$3CuS + 8NO_3^- + 8H^+ \rightarrow 3Cu^{2+} + 3SO_4^{2-} + 8NO + 4H_2O .$$

This equation has minimum integer coefficients, and is the proper balanced ionic equation for this reaction. Since the cations and anions remaining can be combined to give neutral compounds, no additional spectator ions are needed, and we have the final, neutral equation

$$3CuS(s) + 8HNO_3(aq) \rightarrow 3CuSO_4(aq) + 8NO(g) + 4H_2O(l) \ .$$

b. Here $H_2C_2O_4$ is oxalic acid, a weak acid, so it is written in unionized form. The ionic equation is then (omitting H^+)

$$Br_2(aq) + H_2C_2O_4(aq) \rightarrow Br^-(aq) + CO_2(g) \ .$$

Half-reaction couples: $H_2C_2O_4 \rightarrow CO_2$ and $Br_2 \rightarrow Br^-$

Balancing:
step 1: $H_2C_2O_4 \rightarrow 2CO_2$
step 3: $H_2C_2O_4 \rightarrow 2CO_2 + 2H^+$
step 5: $H_2C_2O_4 \rightarrow 2CO_2 + 2H^+ + 2e^-$ (oxidation)

step 1: $Br_2 \qquad\qquad \rightarrow 2Br^-$
step 5: $Br_2 + 2e^- \rightarrow 2Br^-$ (reduction)

Adding (multipliers both 1): $Br_2 + H_2C_2O_4 \rightarrow 2CO_2 + 2H^+ + 2Br^-$

Neutral equation: $Br_2(aq) + H_2C_2O_4(aq) \rightarrow 2CO_2(g) + 2HBr(aq)$

c. Here there are no ionic species present.

Half-reaction couples: $CH_4O \rightarrow CO_2$ and $H_2O_2 \rightarrow H_2O$.

Balancing:
step 2: $CH_4O + H_2O \rightarrow CO_2$
step 3: $CH_4O + H_2O \rightarrow CO_2 + 6H^+$
step 5: $CH_4O + H_2O \rightarrow CO_2 + 6H^+ + 6e^-$ (oxidation)

step 2: $H_2O_2 \qquad\qquad \rightarrow 2H_2O$
step 3: $H_2O_2 + 2H^+ \qquad \rightarrow 2H_2O$
step 5: $H_2O_2 + 2H^+ + 2e^- \rightarrow 2H_2O$ (reduction)

Adding (ox $\times 1$, red $\times 3$):
$$CH_4O + H_2O + 3H_2O_2 + 6H^+ \rightarrow CO_2 + 6H^+ + 6H_2O$$

Cancelling like species: $CH_4O + 3H_2O_2 \rightarrow CO_2 + 5H_2O$

d. This is a *disproportionation* reaction, a key step in the industrial synthesis of nitric acid. In a disproportionation, the same species is both oxidized and reduced. Each of the couples then involves the same reagent.

Ionic form: $NO_2 \rightarrow NO_3^- + NO$

Couples: $NO_2 \rightarrow NO_3^-$ and $NO_2 \rightarrow NO$

Balancing:

   step 2: $NO_2 + \mathbf{H_2O} \rightarrow NO_3^-$
   step 3: $NO_2 + H_2O \rightarrow NO_3^- + \mathbf{2H^+}$
   step 5: $NO_2 + H_2O \rightarrow NO_3^- + 2H^+ + \mathbf{e^-}$ (oxidation)

   step 2: NO_2 $\rightarrow NO + \mathbf{H_2O}$
   step 3: $NO_2 + \mathbf{2H^+}$ $\rightarrow NO + H_2O$
   step 5: $NO_2 + 2H^+ + \mathbf{2e^-}$ $\rightarrow NO + H_2O$ (reduction)

Adding (ox $\times 2$, red $\times 1$):
   $2NO_2 + 2H_2O + NO_2 + 2H^+ \rightarrow 2NO_3^- + 4H^+ + NO + H_2O$

Cancelling/combining like species: $3NO_2 + H_2O \rightarrow 2NO_3^- + 2H^+ + NO$
Neutral equation: $3NO_2(g) + H_2O(l) \rightarrow 2HNO_3(aq) + NO(g)$

e. H_2S is a weak acid, and is not written in ionized form. The specification of HCl as the acid provided identifies the anion spectator as Cl^-.

Ionic form: $H_2S + Cr_2O_7^{2-} \rightarrow S + Cr^{3+}$

Couples: $H_2S \rightarrow S$ and $Cr_2O_7^{2-} \rightarrow Cr^{3+}$

Balancing:

   step 3: $H_2S \rightarrow S + \mathbf{2H^+}$
   step 5: $H_2S \rightarrow S + 2H^+ + \mathbf{2e^-}$ (oxidation)

   step 1: $Cr_2O_7^{2-}$ $\rightarrow \mathbf{2}Cr^{3+}$
   step 2: $Cr_2O_7^{2-}$ $\rightarrow 2Cr^{3+} + \mathbf{7H_2O}$
   step 3: $Cr_2O_7^{2-} + \mathbf{14H^+}$ $\rightarrow 2Cr^{3+} + 7H_2O$
   step 5: $Cr_2O_7^{2-} + 14H^+ + \mathbf{6e^-}$ $\rightarrow 2Cr^{3+} + 7H_2O$ (reduction)

Adding (ox $\times 3$, red $\times 1$):
   $3H_2S + Cr_2O_7^{2-} + 14H^+ \rightarrow 3S + 6H^+ + 2Cr^{3+} + 7H_2O$
Cancelling: $3H_2S + Cr_2O_7^{2-} + 8H^+ \rightarrow 3S + 2Cr^{3+} + 7H_2O$
Neutral: $3H_2S(aq) + K_2Cr_2O_7(aq) + 8HCl(aq) \rightarrow$
                    $3S(s) + 2CrCl_3(aq) + 2KCl(aq) + 7H_2O(l)$

Note that the last step involves adding 8 Cl^- spectators to partner the $8H^+$; these Cl^- then serve to engage the Cr^{3+} product as well as the K^+ spectators.

f. Ionic form: $MnO_4^- + Br^- \rightarrow Mn^{2+} + Br_2$

   Couples: $MnO_4^- \rightarrow Mn^{2+}$ and $Br^- \rightarrow Br_2$

Balancing:

step 2: MnO_4^- $\rightarrow Mn^{2+} + \mathbf{4H_2O}$

step 3: $MnO_4^- + \mathbf{8H^+}$ $\rightarrow Mn^{2+} + 4H_2O$

step 5: $MnO_4^- + \mathbf{8H^+} + \mathbf{5e^-}$ $\rightarrow Mn^{2+} + 4H_2O$ (reduction)

step 1: $\mathbf{2Br^-} \rightarrow Br_2$

step 5: $2Br^- \rightarrow Br_2 + \mathbf{2e^-}$ (oxidation)

Adding (red $\times 2$, ox $\times 5$):

$$2MnO_4^- + 16H^+ + 10Br^- \rightarrow 2Mn^{2+} + 8H_2O + 5Br_2$$

Neutral: $2KMnO_4(aq) + 16HBr(aq) \rightarrow$

$$2MnBr_2(aq) + 5Br_2(aq) + 2KBr(aq) + 8H_2O(l)$$

Note that the only ambient anion is Br^-, which here serves as both spectator counterion and reducing agent.

g. This is an example of *inverse disproportionation*, two reagents separately going to the same product. The product must then appear in both couples.

Ionic form: $IO_3^- + I^- \rightarrow I_3^-$

Couples: $IO_3^- \rightarrow I_3^-$ and $I^- \rightarrow I_3^-$

Balancing:

step 1: $3IO_3^-$ $\rightarrow I_3^-$

step 2: $3IO_3^-$ $\rightarrow I_3^- + \mathbf{9H_2O}$

step 3: $3IO_3^- + \mathbf{18H^+}$ $\rightarrow I_3^- + 9H_2O$

step 5: $3IO_3^- + 18H^+ + \mathbf{16e^-}$ $\rightarrow I_3^- + 9H_2O$ (reduction)

step 1: $3I^- \rightarrow I_3^-$

step 5: $3I^- \rightarrow I_3^- + \mathbf{2e^-}$ (oxidation)

Adding (red $\times 1$, ox $\times 8$): $3IO_3^- + 18H^+ + 24I^- \rightarrow I_3^- + 8I_3^- + 9H_2O$

Combining and dividing by 3: $IO_3^- + 6H^+ + 8I^- \rightarrow 3I_3^- + 3H_2O$

Neutral: $KIO_3(aq) + 3H_2SO_4(aq) + 8KI(aq) \rightarrow$

$$3KI_3(aq) + 3K_2SO_4(aq) + 3H_2O(l)$$

47. The titration data yield n_{Fe}

$$n_{Fe} = (36.79\,mL)(0.2000\,mmol\,MnO_4^-/mL)\left(\frac{10\,mmol\,Fe^{2+}}{2\,mmol\,MnO_4^-}\right) = 36.79\,mmol$$

and $g_{Fe} = (36.79\,mmol)(55.85\,mg/mmol) = 2055\,mg = 2.055\,g$. The percentage Fe in the ore is then %Fe = 2.05 g Fe / 3.3111 g ore $\times$ 100% = 62.05%.

49. See the text key for the balanced dichromate half-reaction; combining this with the

thiosulfate half reaction

$$2S_2O_3^{2-} \rightarrow S_4O_6^{2-} + 2e^-$$

allows us to find

$$n_{K_2Cr_2O_7} = (25\,mL)(0.1\,M\,S_2O_3^{2-})\left(\frac{1\,mmol\,I_2}{2\,mmol\,S_2O_3^{2-}}\right)\left(\frac{1\,mmol\,K_2Cr_2O_7}{3\,mmol\,I_2}\right) = 0.417\,mmol$$

and g $K_2Cr_2O_7$ = (0.417 mmol)(294.2 mg/mmol) = 123 mg = 0.12 g.

51. In the cyanide leaching reaction, the O_2 seems to "disappear"; this is an artifact of not including H_2O or OH^- in the equation. In addition, the $Au(CN)_4^-$ ion, tetracyanoaurate(III), is a type of species, a "transition metal complex", not yet encountered in our redox adventures. Balancing this reaction:

Couples: $Au \rightarrow Au(CN)_4^-$ and $O_2 \rightarrow H_2O$
Balancing:

step 1: $Au + 4CN^- \rightarrow Au(CN)_4^-$
step 5: $Au + 4CN^- \rightarrow Au(CN)_4^- + 3e^-$ (oxidation)

step 2: O_2 $\rightarrow 2H_2O$
step 3: $O_2 + 4H^+$ $\rightarrow 2H_2O$
step 4: $O_2 + 2H_2O$ $\rightarrow 4OH^-$
step 5: $O_2 + 2H_2O + 4e^- \rightarrow 4OH^-$ (reduction)

Adding (ox ×4, red ×3):

$4Au + 16CN^- + 3O_2 + 6H_2O \rightarrow 4Au(CN)_4^- + 12OH^-$

Neutral:

$4Au(s) + 16NaCN(aq) + 3O_2(g) + 6H_2O(l) \rightarrow$
$\qquad\qquad\qquad\qquad 4NaAu(CN)_4(aq) + 12NaOH(aq)$

The second reaction is balanced more simply; the reduction half-reaction is the reverse of the $Au/Au(CN)_4^-$ above, while the oxidation $Zn/Zn(CN)_4^{2-}$ is of similar form, $Zn + 4CN^- \rightarrow Zn(CN)_4^{2-} + 2e^-$, yielding the balanced equation

$3Zn(s) + 2NaAu(CN)_4(aq) + 4NaCN(aq) \rightarrow 3Na_2Zn(CN)_4(aq) + 2Au(s)$

Only this equation is needed for the stoichiometric calculation; it is assumed that NaCN is in excess. We have

$$g\,Zn = 31.1\,g\,Au\left(\frac{1\,mol\,Au}{197.0\,g\,Au}\right)\left(\frac{3\,mol\,Zn}{2\,mol\,Au}\right)\left(\frac{65.39\,g\,Zn}{mol\,Zn}\right) = 15.5\,g$$

53. These reactions are a simple application of the solubility rules given in Section 5.7. The principle consists in examining the ionic form of the equation to discover any combinations of limited solubility. See the text answer key.

55. These reactions are simple applications of the block diagram for classification given in Section 5.7. See the text answer key.

6

Orbitals and Chemical Bonding I
The Valence Bond Model
and Molecular Geometry

Your Chapter 6 GOALS:

- Describe the Heitler-London treatment of H_2, including exchange and orbital overlap
- Be able to sketch bond potential energy curves
- Write valence bond wave functions for various bonds in Lewis structures
- Be able to use hybridization of atomic orbitals to rationalize bonding in molecules
- Express hybrid orbitals as linear combinations of atomic orbitals and sketch hybrid orbitals
- Use VSEPR to predict and rationalize molecular geometries for molecules whose central atom is surrounded by two to six electron pairs
- Differentiate between σ and π bonding
- Use the valence bond model to visualize σ and π bonding in molecules whose Lewis structure contains double and/or triple bonds
- Assess the successes and shortcomings of the valence bond model
- Define a dipole moment
- Use dipole moments to quantify fractional charge separation
- Use the geometry of molecules to assess the overall dipole moments of molecules
- Realize that the valence bond wave functions must be combined with ionic wave functions to describe polar bonds more accurately

Chapter 6 KEY EQUATIONS:

- 6.4, 6.6, 6.8, 6.9
- $\delta = 0.84(1 - \exp[-0.27(\Delta\chi)^{2.4}])$

Overview

   Valence bond (VB) theory is conceptually pleasing and easy to grasp for several reasons. It follows a logical step by using atomic orbitals (AO's) to construct larger valence bond wave functions, and it fits very nicely with Lewis structures. Instead of simply accepting the fact that hydrogen makes one bond (or, has a valence of one, as Lewis would have said), allowing us to write H–H, we are in a position to say that the *exchange* of electrons between $1s$ AO's lowers the electrons' energy by allowing them to bathe in the positive charge of two nuclei instead of just one. The bond that is drawn to connect the H atoms is not just a schematic tool now. It is something real, the result of actual electrons moving in the force field of two nuclei while somehow maintaining their identity as $1s$ electrons. Indeed, the Lewis structures of Chapter 5 are much more credible, as this concept of shared AO's or hybrids of AO's can be extended to many atom molecules such as NH_3 and C_2H_2 and beyond. This individual bond idea is simple: N makes three bonds in NH_3, C makes four bonds in CH_4, etc. The mechanisms that we drew in Chapter 5, in fact, relied very much on this principle. The attacking O^{2-} ion in Equation 5.20, for example, attacks just one of the OH bonds, leaving the other one untouched. You should be developing the background to say that this can't be the exact case in reality, but this compartmentalized view of molecules is exceedingly helpful to organic chemists who use it with great success. We will see more of this in Chapter 10, when we examine the energy changes associated with chemical reactions by looking at *individual* bond energies. (See Table 10.3.)

   It is possible to forget that VB theory is based on quantum mechanics, but it is not just a qualitative patchwork theory. The shapes of the AO's on which VB relies are the result of solving the Schrödinger equation for hydrogen, and the linear combinations that form the hybridized orbitals are legitimate mathematical constructs, the true complexity of which has not been shown (see Exercise 1). If VB theory were not based somewhat on the truth, which is quantum mechanics, it could not possibly have the success that it does. One resounding success is seen in Exercise 12, where we find that the Lewis acidity of BF_3 can be attributed to an empty, unhybridized p orbital on the sp^2 hybridized B atom of BF_3. Section 6.6 highlights the failings of VB theory, which stem from the theory's reliance on the electron pair. Electrons do not always appear in pairs; O_2 and NO prove this. The molecular orbital theory of Chapter 7 tries to deal with this matter.

Exercise 2 explores the critical nature that symmetry plays in chemical behavior. Only orbitals of like symmetry have non-zero overlap. This appears again in Chapter 7 and in many other areas of chemistry. Exercise 3 should be read every morning while you are eating breakfast. It is chock-full of information, covering a wide variety of topics. The Coulomb potential should be familiar to you and the inverse-power dependence will be seen again in Chapter 14. (See Equations 14.17 and 14.18.)

Atomic hybridization is not a verifiable fact. It is a scheme that chemists have devised in response to the knowledge that has been gained about the actual shape of molecules and can be used, for example, to predict the outcome of an untried reaction or to guide the design of drugs. To examine this topic, let's look at acetic acid in Exercise 10, where you are asked to consider the non-terminal O. Here, however, we'll consider the carbonyl oxygen. The σ bond of the carbonyl can be described as being formed between the carbon sp^2 orbital and an unhybridized p orbital *or* an sp^2 orbital on the O atom. In either case, the π bond can be explained by the overlap of unhybridized p orbitals on C and O. If hybridization were "true", an alternate explanation wouldn't suffice. We invoke hybridization only when it is needed, a sure sign that it is not a general explanation. Thus, interpretation and philosophical leanings can influence how you view bonding. In the case just cited, there is no harm, but at times it seems that some chemists rely too heavily on the picture that hybridization gives us *with misleading results*. The rabbit ears representation of the lone pairs on water or the articulated lone pair lobe on NH_3 (see Section 6.5 for the structures) are picturesque; they are not, however, true. Their inclusion is to remind you that sp^3 hybridization is being used to rationalize the bonding in these two molecules. Also, they provide conceptually convenient places to start a mechanism, if, for example, water should decide to act as a Lewis base and donate an electron pair.

Several points argue strongly against the validity of lone pairs existing in such an arrangement and the validity of hybridization in general. First, at the end of Section 6.2 it is stated that a spherical charge density, which is present before hybridization, can't give rise to a non-spherical charge density after hybridization. The rabbit ears are clearly non-spherical. Note that we are not saying that the water molecule is spherical. It is not. The distribution of the lone pairs, however, is spherical. Second, the Heisenberg uncertainty principle and the wave nature of electrons prevent localization of electron density. The electrons simply will not huddle in such a small volume and remain localized there. Everything that was presented in Chapters 2 and 3 says this is *impossible*. They will extend over the molecule as much as they can in an effort to lower their energy. Third, there is no physical basis for such a placement of the electrons. Charge separation occurs when the electrons are drawn one way or the other. To what are they drawn? *Positive charge*. No positive charge exists away from the nucleus and the electrons would not decide to simply take a sojourn to the molecule's periphery and stop, which is what the rabbit ears picture posits. Fourth, Potts and Price (1972) used spectroscopy to show that the valence electrons in CH_4, that is, excluding the $1s$ electrons of carbon, exist in *two* different electronic environments. If the hybridization picture were 100 % correct, the valence electrons of CH_4 would all exist in the *identical* environment – in four sp^3 orbitals. We argue these points to make certain that you begin to develop the rational skepticism of which we have spoken. *University*

Chemistry provides you with *models* for chemical behavior, schematic pictures of molecules, etc.. Be very careful of the meaning and significance that you assign to these most mysterious creatures that we call molecules.

This concern over the representation of lone pairs is not purely academic. It has a strong impact on the way you should view the overall dipole moments of molecules (Exercises 18 through 24). To repeat from the text, lone pairs do not contribute to the overall dipole moment. To say that they do, ascribes too much validity to them. What, then, *is* responsible for an overall dipole moment? *A net dipole results when there is an imbalance of bond dipoles.* The N–H bonds of ammonia are not symmetrically distributed, nor are the O–H bonds in water symmetrically distributed. This asymmetrical distribution, not the lone pairs, gives rise to the overall dipole moment. The charge imbalance of an N–H or O–H bond is real. It is *not* schematic, as is the articulated picture of a lone pair. For this reason, you must focus on the real concepts when deciding dipole matters. The given answer for Exercise 21 is sufficient, but for those who like a challenge, you can derive a similar formula for molecules such as NH_3. The result is $\mu = 3\mu_b\cos\alpha$, where α is the angle that each individual bond dipole makes with the so-called C_3 axis of the general molecule AB_3. The C_3 axis is the axis around which you could rotate the molecule 120° and still have the molecule look identical.

It is customary to omit lone pairs when they are not important. When looking at any of the Lewis structures that are presented for Exercises 13 and 15, don't be alarmed that many of the terminal halides or oxygen atoms are missing their lone pairs. It would not, however, be acceptable to omit the two lone pairs, for example, that are placed on the central Xe atom of XeF_4 in Exercise 14(i).

1. See the text answer key. The normalization given in the text and the answer key is approximate, neglecting the effect of overlapping of the AO's. The rigorously normalized Valence Bond wave function is given by

$$\Psi(\mathbf{r}_1,\mathbf{r}_2) = [2 + 2S^2]^{-\frac{1}{2}}[\psi_a(\mathbf{r}_1)\psi_b(\mathbf{r}_2) + \psi_a(\mathbf{r}_2)\psi_b(\mathbf{r}_1)]$$

where $S = \int \psi_a\psi_b\, d\tau$ is the orbital overlap defined in Equation 6.3. Thus simply dividing the sum of orbital products by $\sqrt{2}$ is valid at large r, where the Heitler-London *ansatz* is also exact. In either case, when the energy is calculated according to

$$E = \frac{\int \int \Psi \hat{H} \Psi \, d\tau_1 \, d\tau_2}{\int \int \Psi^2 \, d\tau_1 \, d\tau_2}$$

the normalization constant cancels out. For LiH, the full wave function should include a contribution from the ionic structure Li^+H^-, in which both valence electrons reside on H, and may be written as

$$\Psi = a\,[2 + 2S^2]^{-\frac{1}{2}}[2s_{Li}(\mathbf{r}_1)1s_H(\mathbf{r}_2) + 2s_{Li}(\mathbf{r}_2)1s_H(\mathbf{r}_1)] + b\,1s_H(\mathbf{r}_1)1s_H(\mathbf{r}_2)$$

where a and b are coefficients such that $a^2 + b^2 = 1$. See the discussion in Section 6.5 and Exercise 25.

3. The bond energies of H_2 and NaCl are similar, but $r_e(NaCl) > r_e(H_2)$, and the attractive limb of the potential energy curve for NaCl has a much longer range. The graph on the next page compares the potential energy curves on the same distance and energy scales. (The decay of the covalent interaction is roughly exponential, $-Be^{-br}$, giving way to an inverse-power

dependence, $-C/r^6$, at very large r. The ionic curve, on the other hand, decays as the Coulomb potential, $-e^2/r$, until $r = r_x$ is reached.) At the crossing point r_x the ionic curve undergoes a sharp change in slope (not shown on the graph), while the covalent curve smoothly approaches zero at large r. The origin of the bonding in NaCl is electrostatic, explicable by classical concepts without the need for orbital overlap, while in H_2 bond formation is strictly a quantum-mechanical phenomenon, arising from a nonclassical exchange energy accompanied by orbital overlap. (In Chapter 7, a deeper explanation of covalent bonding in terms of delocalization of electron-waves will be given.)

5. From Table 3.2 we have the orbital functions

$$2s = \frac{1}{4\sqrt{2\pi}} a_0^{-3/2} (\rho - 2) e^{-\rho/2} \qquad 2p_x = \frac{1}{4\sqrt{2\pi}} a_0^{-3/2} \rho e^{-\rho/2} \sin\theta \cos\phi$$

We have reversed the sign of the 2s orbital as suggested; this yields a positive amplitude outside the 2s radial node. We form the sp hybrid orbitals as in Equation 6.6:

$$sp_1 = \frac{1}{\sqrt{2}}(2s + 2p_x) = \frac{1}{8\sqrt{\pi}a_0^{3/2}} (\rho - 2 + \rho\sin\theta\cos\phi) e^{-\rho/2}$$

$$sp_2 = \frac{1}{\sqrt{2}}(2s - 2p_x) = \frac{1}{8\sqrt{\pi}a_0^{3/2}} (\rho - 2 - \rho\sin\theta\cos\phi) e^{-\rho/2}$$

where we have factored out the common normalization constant and exponential. Along the x axis, $\rho = |x|/a_0$, $\theta = 90°$, and $\phi = 0°(180°)$ for $x >(<) 0$; then the product $\sin\theta\cos\phi = +1$ for $x > 0$ and -1 for $x < 0$, and we have

$$sp_1 = \frac{1}{8\sqrt{\pi}a_0^{3/2}} (2\rho - 2) e^{-\rho/2}, \quad x > 0;$$

$$= \frac{1}{8\sqrt{\pi}a_0^{3/2}} (-2) e^{-\rho/2}, \qquad x < 0,$$

From the $x > 0$ form, we find a node at $2\rho - 2 = 0$, or $r = a_0$. To find the position of the maximum, we set $d(sp_1)/d\rho = 0$ (as in Exercise 3.11). In setting the derivative to zero, we can ignore the normalization constant, since it will cancel. We then have

$$\frac{d(sp_1)}{d\rho} \propto \frac{d}{d\rho}[(2\rho - 2)e^{-\rho/2}]$$

$$= [2 - (2\rho - 2)/2]e^{-\rho/2}$$

$$= (3 - \rho)e^{-\rho/2} = 0$$

yielding a maximum at $r = 3a_0$. For $x < 0$ the hybrid shows no zeroes or maxima, and simply dies away exponentially, while at $x = 0$ it shows a cusp due to the 2s contribution. A plot of the hybrid orbital amplitude along x is shown at the right. This analysis shows quantitatively how the component orbitals interfere constructively for $x >$

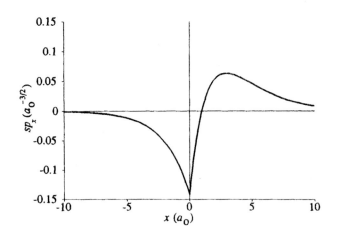

0 and destructively for $x < 0$, as illustrated here and in "3-D" by the boundary surfaces of Figure 6.5. The node along the x axis found above is the apex of a nodal surface, whose form is given by $x = 2a_0 - r$, yielding a paraboloid of revolution around the x axis.

7. The new aspect in the Exercise is the presence of one or more double bonds in these

molecules; the π component of each double bond is ignored. We construct the table

Species	central atom	# σ bonds on cent at	# lone pr on cent at	# e pr (σ + lp)	hybrid.	bond angles
a) CO_2	C	2	0	2	sp	$180°$
b) NO_2^-	N	2	1	3	sp^2	$\leq 120°$
c) NO_3^-	N	3	0	3	sp^2	$120°$
d) SO_3	S	3	0	3	sp^2	$120°$
e) O_3	O	2	1	3	sp^2	$\leq 120°$
f) ClO_3^-	Cl	3	1	4	sp^3	$\leq 109°$

9. The valence bond overlaps are shown at the right. The lone pairs will be ensconced in atomic O $2s$ and $2p$ orbitals, as in Chapter 3. The O atom can also be regarded as being sp^2-hybridized; this changes the picture of the σ overlap to involve two overlapping hybrids, as in ethylene, Figure 6.7, and the lone pairs then occupy the remaining O atom hybrids instead of the unhybridized AO's. Although the valence bond overlap picture gives the impression that most of the electron-wave amplitude is on the C atom, this is not the case. The O atom orbital has a higher amplitude that compensates for its smaller spatial extent; further, the picture neglects the ionic contribution to the bond, which also enhances the density on O. (See the discussion of Exercise 1 above, and Section 6.5 in the text.)

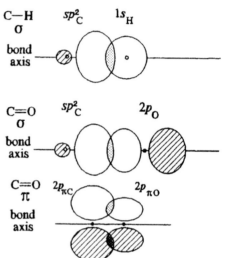

11. See the text answer key, and Exercises 5.10 and 5.14 for Lewis structures. As in Exercise 7, we avoid expanding the octet for S in H_2SO_4 and P in H_3PO_4 for simplicity; then HNO_3 alone contains a double bond and hence a π bond along with sp^2 N hybridization. As in Exercise 9, we leave the end atoms unhybridized, since the arrangement of their electrons does not affect the nuclear geometry. Note how different are the Group V oxyacids HNO_3 and H_3PO_4: the larger P atom cannot make strong π bonds, while at the same time it can more readily accommodate another bonded group. Some similarity between the two can be recognized by writing H_3PO_4 as $HPO_3 \cdot H_2O$, a Lewis acid-base adduct; water, acting as a Lewis base, can be viewed as causing a weak π bond in HPO_3 to cleave in favor of a new P—O bond. This motif enables the energy-bearing polyphosphate structures in the molecules of life (see Chapter 18.)

13. See text answer key for
    geometries and bond angles.
    Geometrical sketches
    conforming to the format of
    Figure 6.11 are given on the
    right.

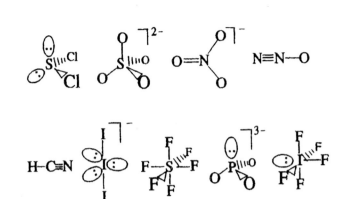

15. Here more than one atom is a bond-
    angle center. See the text key for bond
    angles and hybridization. Geometrical
    sketches are shown at the right; for
    simplicity bonds on different atomic
    centers are shown as lying in a
    common plane if possible, although that
    might not represent the most stable
    conformation. Conformational issues
    are considered in Chapter 18;
    molecular model kits can be very
    helpful for these aspects of structure.

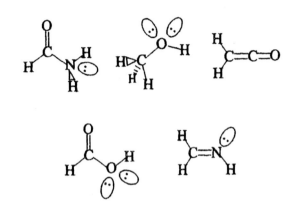

17. See text answer key. It is generally assumed in attempting to draw Lewis structures of
    free-radical triatomics that the unpaired electron resides on the central atom. The
    principles behind finding the geometry of the ions are the same as explored in the
    previous four Exercises.

19. Using dipole and bond length data from Table 6.2, we have

$$\delta_{KBr} = \frac{\mu}{er_e} = \frac{10.41\ D}{(4.8)(2.82)\ D} = 0.77$$

$$\delta_{HBr} = \frac{0.82\ D}{(4.8)(1.41)\ D} = 0.12$$

while Equation 6.9 yields

$$\delta_{KBr} = 0.84\ \{1 - \exp[-0.27(2.2)^{2.4}]\} = 0.70$$

$$\delta_{HBr} = 0.84\ \{1 - \exp[-0.27(0.8)^{2.4}]\} = 0.12$$

The Lewis structures with dipolar labeling are: $K^+[:\overset{..}{Br}:]^-$ and $H-\overset{..}{\underset{..}{Br}}:$.

21. As illustrated on the next page using H_2O as an example, the bond dipoles are added

vectorially to yield the overall net dipole of the molecule. For molecules of the AB_2 type, each bond dipole μ_b has the same length, so that when they are placed head-to-tail for vector addition, they form an isosceles triangle with the overall dipole μ. The base angle of this triangle may be deduced from plane geometry to be $\theta/2$, as shown. When a perpendicular is dropped from the apex of this triangle,

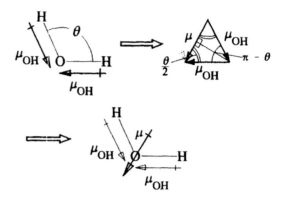

μ is divided into two equal pieces, each of which is a leg of a right triangle of hypothenuse μ_b and adjacent angle $\theta/2$. By the definition of the cosine, we have $\mu/2 = \mu_b\cos(\theta/2)$ or $\mu = 2\mu_b\cos(\theta/2)$, QED. Note that μ bisects the bond angle of the triatomic, as illustrated. Of course, only the overall dipole μ is measurable, so this enables us to find μ_b, $\mu_b = \mu/[2\cos(\theta/2)]$, as well as the charge separation $\delta = \mu_b/(er_e)$, where r_e is the bond length. Carrying through the calculation for H_2O using data from Table 6.1 and the bond angle given in Chapter 5:

$$\mu_b = \frac{\mu}{2\cos(\theta/2)} = \frac{1.85\,D}{2\cos(104.5^\circ/2)} = 1.51\,D$$

$$\delta = \frac{\mu_b}{er_e} = \frac{1.51\,D}{(4.8)(0.96)\,D} \approx 0.33$$

For the other molecules, ths calculation is analogous; see the text answer key. For NO_2 and H_2S, the anomalous bond angles given in Table 6.1 were used, while for SO_2, the VSEPR estimate was used. Using VSEPR is usually adequate for this purpose; for example, the experimental bond angle for SO_2 is 118°, while the VSEPR value is $(\leq)$ 120°, but these yield the same μ_b and δ to 2-3S. Finally, we note that the partial charge on each H in H_2O is $+0.33$, while that on O is -0.66, -0.33 coming from each bond dipole.

23. The molecules HOCl, H_2O, HOF, and Cl_2O are expected to have similar bond angles, $\sim 105^\circ$, based on VSEPR. Further, each of the bond dipoles is

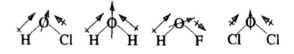

expected to point toward the more electronegative O atom in all cases except O—F. Based on the Pauling correlation (see Section 6.5), we expect μ_b to increase with $\Delta\chi$, and thus that $\mu_b(O—Cl) \approx -\mu_b(O—F) < \mu_b(O—H)$. The situation is illustrated qualitatively above; the estimated resultant dipoles are the dashed arrows. The ordering $\mu(Cl_2O) < \mu(HOCl) < \mu(H_2O)$ is unambiguous, owing to $\mu_b(O—Cl) < \mu_b(O—H)$. Since the bond angle is greater than 90°, we expect that $\mu_b(O—F)$ will reinforce $\mu_b(O—H)$ to a greater extent than $\mu_b(O—Cl)$ does, and hence that $\mu(HOCl) < \mu(HOF) < \mu(H_2O)$, giving an overall order $\mu(Cl_2O) < \mu(HOCl) < \mu(HOF) < \mu(H_2O)$. Aside from H_2O, these dipole moments have not yet been measured.

25. See the text answer key and the discussions in Section 6.5 and in Exercise 1 above. The calculation of δ can be made either from $\delta = \mu/(er_e)$ with the aid of Table 6.1, or by using Equation 6.9, both of which are discussed in Exercise 19 above.

7

Orbitals and Chemical Bonding II
The Molecular Orbital Model
and Molecular Energy Levels

Your Chapter 7 GOALS:

- Understand the reason why H_2^+ is the paradigm for MO theory
- Note that the extension of the H_2^+ MO's to other diatomic molecules is analogous to the extension of hydrogen's AO's to other atoms
- Draw the MO's that result from the linear combination of atomic orbitals
- Attribute covalent bond formation to delocalization
- Be able to compare and contrast the VB and MO approaches to bonding
- Write MO electron configurations for molecules
- State Mulliken's Rules
- Assess the successes and shortcomings of MO theory
- Correlate the nodal structure of MO's with the nodal structure of the Particle in a Box wave functions
- Be able to relate bond order to bond strength
- Extend the homonuclear diatomic MO picture to heteronuclear diatomic molecules and linear triatomic dihydrides
- Use MO theory to rationalize reaction mechanisms and explain the existence of resonance structures
- Couple VB and MO theory to describe compounds with σ and π bonding

Chapter 7 KEY EQUATIONS:

- 7.1, 7.2, 7.3, 7.4

Overview

Chapter 7 is another "at bat" for chemists as we try to describe bonding in molecules. The VB theory of Chapter 6 is still widely used by many chemists, but there must be a way to explain why O_2 is paramagnetic, an approach that easily handles a molecule like NO, and an explanation as to why CH_4 has two different kinds of electrons in its valence orbitals, not one. (See the Chapter 6 introduction for the CH_4 discussion.) The LCAO-MO approach of Mulliken manages to do all three of these things. The resolution of the O_2 dilemma is particularly impressive because MO theory correctly predicts the magnetic properties of O_2, whereas VB theory does not. Additionally, MO theory does not require the electrons to retain their identity as AO's once the molecule is formed. This seems to be a more natural state of affairs, given, for example, that a hydrogen electron does not think. It responds to forces that act on it, and since each proton of H_2 can pull with equal strength, there is no reason to expect an electron to stay with its initial H atom in an H_2 molecule. (VB theory allows for this, too, via the *exchange* energy, but VB theory always needs some retention of AO's in the molecule.) Speaking of baseball, we rate MO theory as a triple and VB theory as a double. Perhaps, there's a chemical Hank Aaron (aka The Hammer) in the audience who will one day round the bases with a unifying theory. We still don't have one!

A new theory is truly successful when it *predicts* many things successfully. After the fact rationalizations, such as hybridization, are not predictions. In this regard, MO theory is much closer to a "from first principles" theory than VB theory is. Such a case is clearly demonstrated at the start of Section 7.2. Note the effortless ease with which MO theory handles the progression from H_2^+ to He_2, making a *successful* prediction about the stability of H_2^+ by utilizing the concept of bond order within a delocalized MO. The Lewis-VB theory (recall that Pauling injected orbitals into Lewis structures, see the end of Section 6.1) stumbles mightily with H_2^+, as shown in Exercise 1, where the Lewis picture of H_2^+ predicts an unstable molecule.

Solve Exercise 2 by setting dK_1 equal to $d(h^2/8ma^2)$, noting that h and m are constants. The united atom of Exercise 4 is He. Exercises 6 through 13 require the use of Figures 7.8 and 7.9 with the bond orders given by Equation 7.3. Bond orders are the key factor in decisions regarding bond lengthening or shortening, bond strengthening or weakening, etc. upon addition or removal of electrons. Just remember that the higher the bond order is, the shorter and stronger is the bond, as shown in Figure 7.9. Note that Exercise 8 is different than the others in this group because it is an $n = 3$ homonuclear diatomic molecule Al_2. Predicting that actual order of the σ_p and π_p levels is risky business, and in this case you can rely on the first sentence that follows Example 7.2, which predicts no shift in the order of these two MO's. Further, the good news is that, even if the order is the opposite of what one expects, the number of bonding and anti-bonding electrons does not change no matter how the σ_p and π_p levels are arranged. Since bond orders depend solely on the *numbers* of these electrons, the importance of this ordering is minimized except in some cases regarding frontier orbitals. See below. Before we move on, note the failings of the VB picture with respect to Al_2. An MO bond order of one suggests a single σ bond in VB theory, involving the $3p_\sigma$ $(3p_z)$ orbitals on each atom. The $3s$ orbitals are filled and non-bonding in this case. (sp or sp^2 hybridization produces unpaired electrons here, which hybridization can't ever do, so that's not an option.) Thus, the VB prediction for Al_2 is that all of the electrons are paired, which can't be true because we know that Al_2 is paramagnetic.

Exercises 14 through 18 and 24 rely on Figure 7.11, with an energy level switch of the σ_p and π_p MO's for OF^-, ClO, and ICl. Note, again, that this does not affect the bond orders. Even in the case of Exercise 24, which is a frontier orbital question, this switch has no impact on the answer. [Only for a case such as CS, which has a fleeting existence, or BCl perhaps (both mix *different n* levels), would a question about the frontier orbital be problematic without a hint or two. Fill in the MO scheme of Figure 7.11 to see why!] To expand on Exercise 24, ICl is the oxidizing agent, and its LUMO, the σ_p^* orbital, interacts with the $4s$ orbital of K. Since ICl's LUMO has greater amplitude on I, more successful overlap with K's $4s$ orbital is expected at the I end of the molecule and a larger yield of KI is predicted. For Exercise 22 the Arrhenius reaction is described by the n_{2p} HOMO on OH^- donating an electron pair into the "LUMO" of H^+, the hydrogen $1s$ orbital.

Exercises 25 and 28 exhibit one of the shortcomings of MO theory that is discussed in Section 7.5 of the text. The process of geometry determination for the linear triatomic dihydrides of Section 7.3 is almost certainly new to you, and its difficulty is not your imagination. It *is* difficult. It seems even more difficult (maybe, unnecessary) when you consider the ease with which you can solve Exercise 25 with VSEPR theory. CH_2 is quickly drawn as

$$H - \overset{..}{C} - H,$$

and Figure 6.11 indicates, correctly, a bent geometry. This same process in MO theory first requires constructing the MO's as in Figure 7.16, followed by the bending diagram (also known as a Walsh diagram) that is shown in Figure 7.17. Even though these two steps have been done for you, understanding them is not simple. It is complicated further by what might appear to you to be a sea of labels, which even change upon bending. All that we can say is that this is not pointless labeling. Both bonding between orbitals and electronic transitions depend very strongly on the symmetry of the orbitals involved. These labels indicate the symmetry of these orbitals and are very helpful to experienced chemists when descriptions of chemical reactions or transitions are attempted. To answer Exercises 25 and 28 you should focus on the number of electrons that decrease in energy upon bending as opposed to the number of electrons whose energy increases upon bending. The full solution in this manual to Exercise 25 shows that in the predicted or actual configuration, a bent geometry is expected. For Exercise 28 use the MO's that are shown for Exercise 27, filling the lowest energy MO with the two available electrons.

The MO's of H_2S for Exercise 26 are similar to those of H_2O, but they are larger and more diffuse. Also, the bonding is more covalent in H_2S than in H_2O due to the smaller electronegativity difference between S and H. When inspecting the MO's of Exercise 27, two of which you should have obtained without error (σ_1 and σ_3), note that σ_2 is "missing" an AO. You should be able to see easily that the one node that σ_2 possesses coincides with the first excited state of the Particle in a Box. Arriving at the exact structure of σ_2, however, is difficult at this level of presentation.

Sections 7.4 and 7.6 as well as Exercises 29 through 32 allow us to present the view of bonding that is shared by many chemists today. The marriage of the VB and MO theories relies on two main concepts: a σ-bonded framework and delocalized π orbitals. See Example 7.5 and Figure 7.20. In Exercises 29 and 30 the N and S atoms are sp^2 hybridized, forming a σ-bonded framework with the O atoms. The unhybridized p orbitals on all of the atoms overlap to form the delocalized π orbital. Likewise, the C of Exercise 31 is sp^2 hybridized, but be careful not to include the H atom in the delocalized π orbital since it has no p orbitals. Lastly, for Exercise 32 note that the σ-bonded framework is present in both cases. The VB model, however, predicts that two localized C=O bonds exist at right angles to each other, whereas the MO picture allows for two delocalized π orbitals, still at right angles to each other, that spread over the entire O$-$C$-$O molecule.

1. The two arrangements of charges in H_2^+ are illustrated at the right. The total potential energy is given in Figure 7.1 as the sum of the two electron-nucleus attractions and the nucleus-nucleus repulsion,

$$V = -\frac{e^2}{r_a} - \frac{e^2}{r_b} + \frac{e^2}{r} = 14.40 \text{ eV Å} \left(-\frac{1}{r_a} - \frac{1}{r_b} + \frac{1}{r} \right)$$

We have again invoked the "engineering" formula of Chapter 1. The sum of the electron-proton distances is constant, $r_a + r_b = r$. In case (a), we have $r_a = r_b$, implying $r_a = r_b = r/2 = (1.06 \text{ Å})/2 = 0.53 \text{ Å}$. In case (b) $r_a = r_b - 0.25\text{Å}$, implying $2r_b - 0.25\text{Å} = 1.06\text{Å}$, or $r_b = (1.06 + 0.25)/2 = 0.65_5 \text{ Å}$ and $r_a = 1.06 - 0.65_5 = 0.40_5 \text{ Å}$. The subscripted digits are uncertain, but are not dropped to avoid rounding errors in differences between potential terms. The potential energies are then

   a. $V = 14.40 \text{ eV Å} \left(-\frac{2}{0.53 \text{ Å}} + \frac{1}{1.06 \text{ Å}} \right) = -40.8 \text{ eV } (-6.53 \times 10^{-18} \text{ J})$

   b. $V = 14.40 \text{ eV Å} \left(-\frac{1}{0.405 \text{ Å}} - \frac{1}{0.655 \text{ Å}} + \frac{1}{1.06 \text{ Å}} \right) = -44.0 \text{ eV } (-7.04 \times 10^{-18} \text{ J})$

Thus the potential energy for the unsymmetrical configuration is lower by 3.2 eV, an amount that exceeds the bond energy of H_2^+. From the Virial Theorem ($E = V/2$) we expect that the total energy E will also be lower, and hence the unsymmetrical arrangement of charges is *more* stable. However, when he was very young Richard Feynman proved that there are no absolutely stable configurations of a system of charged particles, except when all the negative charges "sit on" the positives and neutralize them. The further we move the electron toward one nucleus, the lower V and E will be. Thus Lewis's dot structure is not stable, and it cannot represent a chemical bond in a physical sense. By differentiating V above with respect to r_a for fixed r ($r_b = r_a - r$), you can show that Lewis's centered electron actually represents a *maximum* in V, i.e., a point of *metastable* equilibrium. Quantum mechanics offers a way out of this "Coulomb impasse" by allowing the electron to act as a wave that "covers" a range of charge configurations at once. As Figure 7.5 shows, the wave still has its highest amplitude near the nuclei, rather than between them.

3. As the two nuclei in H_2^+ are pushed together, they become equivalent to a He nucleus, $Z = 2$, and we have a He^+ cation. We therefore expect that Z_{eff} will vary between 1, its value when H_2^+ is dissociated into $H + H^+$, and 2: $1 \leq Z_{eff} \leq 2$. We note that the so-called *variational* value of Z_{eff}, that which gives the lowest energy at the minimum of the bond potential energy curve, lies between these limits. Using a relationship similar to that used for H_2 (see the solution to Exercise 6.1), the energy may be calculated from

$$E = \frac{\int \psi \hat{H} \psi \, d\tau}{\int \psi^2 \, d\tau}$$

where the properly-normalized bonding MO wave function is

$$\psi = \sigma_{1s} = \frac{1}{\sqrt{2+2S}} [1s_a + 1s_b]$$

and S is the overlap integral, $S = \int 1s_a 1s_b \, d\tau$. In the text the MO's are given in unnormalized form, but as mentioned in the Exercise 6.1 solution, the normalization constant cancels when the energy is computed.

5. For this Exercise it is helpful to make sketches of the overlapping orbitals, as shown at the right. We have used the convention that the positive lobe of a p_z orbital lies nearest to the neighboring atom, and we match the signs of orbital lobes (indicated as before by shading) on adjacent atoms where possible to give constructive ("bonding") overlap. From the sketches it is clear that cases (a) and (b) produce cancelling

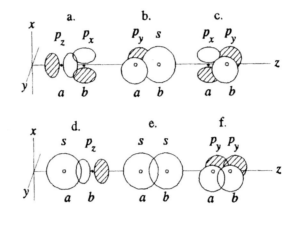

contributions to $S = \int \psi_a \psi_b \, d\tau$ arising from opposite sides of the y-z and x-z planes, respectively. For case (c), considering either the y-z or x-z planes yields cancellation. Thus $S = 0$ for cases (a), (b), and (c). For cases (d), (e), and (f), any dividing plane containing the z axis produces equal, but not opposite, contributions to S, and hence $S \neq 0$ for these cases. In case (d), S does become zero in the united atom limit. Cases (d) and (e) correspond to σ overlap, case (f) to π.

7. With only 4 electrons in the valence shell, only the σ_{2s} orbitals are occupied in the ground state of Be_2, as shown at the right, leading to an electron configuration
    $$(\sigma_{2s})^2 (\sigma_{2s}^*)^2$$
    and a bond order BO $= \frac{1}{2}(2-2) = 0$. We thus predict that no bonding should occur between two Be atoms; the very weak "bond" observed is consistent with this conclusion, and arises from

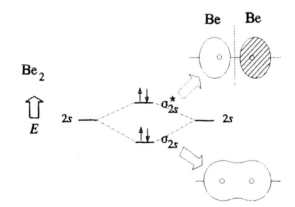

s-p mixing effects that are beyond our level here. The MO description of Be_2 at our simple level is analogous to that for He_2. The σ MOs are of the form shown in Figures 7.6 and 7.7, and are also illustrated here.

9. We may assume the ions are in their ground states; although initially they may be formed with some probability in excited states, these usually radiate and reach the ground state very quickly. The ground state of the ion results from removing an electron from the highest-energy occupied MO (HOMO) of the neutral species. The electron configurations and bond orders of neutral N_2 and O_2 are given in Section 7.2, from which we find:

N_2^+: $(\sigma_{2s})^2(\sigma_{2s}*)^2(\pi_{2p})^4(\sigma_{2p})^1$;   BO $= \frac{1}{2}(7-2) = 2\frac{1}{2}$;

O_2^+: $(\sigma_{2s})^2(\sigma_{2s}*)^2(\sigma_{2p})^2(\pi_{2p})^4(\pi_{2p}*)^1$;   BO $= \frac{1}{2}(8-3) = 2\frac{1}{2}$.

Note that the energy reversal of the σ_{2p} and π_{2p} orbitals between N_2 and O_2 has no effect on the BO calculation. For $N_2 \rightarrow N_2^+$, the BO therefore decreases from 3 to $2\frac{1}{2}$, implying that the bond is weakened ($D_e \downarrow$) and lengthened ($r_e \uparrow$) as a result of ionization. Just the reverse is the case for $O_2 \rightarrow O_2^+$; BO increases from 2 to $2\frac{1}{2}$, and the bond is strengthened ($D_e \uparrow$) and shortened ($r_e \downarrow$). In N_2 a bonding electron is lost; in O_2 an antibonding one. The following experimental values of D_e and r_e back up this MO deduction:

Species	N_2	N_2^+	O_2	O_2^+
D_e (eV)	9.76	8.72	5.11	6.70
r_e (Å)	1.10	1.12	1.21	1.12

11. The electron configurations of the negative ions and the bond orders of ions and their neutral precursors are

		BO$_{anion}$	BO$_{neutral}$
C_2^-:	$(\sigma_{2s})^2(\sigma_{2s}*)^2(\pi_{2p})^4(\sigma_{2p})^1$	$\frac{1}{2}(7-2) = 2\frac{1}{2}$	2
N_2^-:	$(\sigma_{2s})^2(\sigma_{2s}*)^2(\pi_{2p})^4(\sigma_{2p})^2(\pi_{2p}*)^1$	$\frac{1}{2}(8-3) = 2\frac{1}{2}$	3
O_2^-:	$(\sigma_{2s})^2(\sigma_{2s}*)^2(\sigma_{2p})^2(\pi_{2p})^4(\pi_{2p}*)^3$	$\frac{1}{2}(8-5) = 1\frac{1}{2}$	2

Only for C_2 does the BO increase when an extra electron is added, thereby stabilizing the anion. MO boundary surface sketches of the newly-occupied orbitals (see Figures 7.6 and 7.7) are shown at the right; note the relative sizes of the MOs.

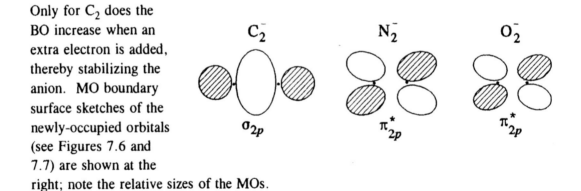

13. If Ca is present as Ca^{2+} in CaC_2, C_2^{2-} (acetylide ion) must exist; its electron configuration is predicted to be

   C_2^{2-}: $(\sigma_{2s})^2(\sigma_{2s}*)^2(\pi_{2p})^4(\sigma_{2p})^2$;   BO $= \frac{1}{2}(8-2) = 3$.

It is isoelectronic with N_2.

15. ClO is a heteronuclear diatomic containing atoms from different periods; to build MO's we must interfere the valence $3s$ and $3p$ orbitals on Cl with the $2s$ and $2p$ on O. χ_O (3.5) is only slightly greater than χ_{Cl} (3.2), implying that the valence atomic orbital energies are not greatly different. A schematic energy-level diagram is shown at the right. The orbitals are labeled simply with s and p subscripts; σ_s is an abbreviation of

σ_{3s2s}. With 13 valence electrons ClO is analogous to O_2^- (Exercise 11), having the electron configuration and bond order

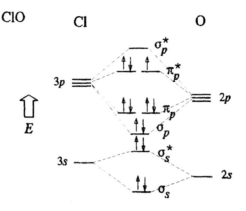

ClO: $(\sigma_s)^2(\sigma_s{}^*)^2(\sigma_p)^2(\pi_p)^4(\pi_p{}^*)^3$;
$$BO = \tfrac{1}{2}(8-5) = 1\tfrac{1}{2}.$$

The bond in ClO consists of a σ and half a π bond; ClO is expected to have a stronger, shorter bond than a typical Cl single bond as in Cl_2. The following data bear this out:

Species	Cl_2	ClO
D_e (eV)	2.48	2.75
r_e (Å)	1.99	1.57

17. The electron configurations of the negative ions and the bond orders of ions and their neutral precursors are

		BO_{anion}	$BO_{neutral}$
BC^-:	$(\sigma_{2s})^2(\sigma_{2s}{}^*)^2(\pi_{2p})^4$	$\tfrac{1}{2}(6-2) = 2$	$1\tfrac{1}{2}$
CO^-:	$(\sigma_{2s})^2(\sigma_{2s}{}^*)^2(\pi_{2p})^4(\sigma_{2p})^2(\pi_{2p}{}^*)^1$	$\tfrac{1}{2}(8-3) = 2\tfrac{1}{2}$	3
ClO^-:	$(\sigma_s)^2(\sigma_s{}^*)^2(\sigma_p)^2(\pi_p)^4(\pi_p{}^*)^4$	$\tfrac{1}{2}(8-6) = 1$	$1\tfrac{1}{2}$

Only for BC is the extra electron added to a bonding MO, shortening and strengthening its bond. The bonds in CO and ClO are weakened and lengthened by the added antibonding e^-. Boundary surface sketches for the added electron are shown at the right (see also Figure 7.12).

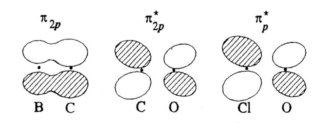

19. An energy level diagram and the corresponding MO boundary surface sketches for HCl are shown at the right. The corresponding electron configuation is

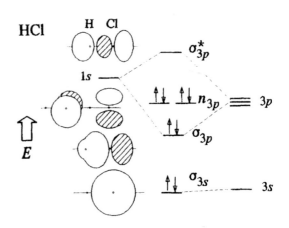

$$(\sigma_{3s})^2(\sigma_{3p})^2(n_{3p})^4$$

and the bond order is 1, arising only from the σ_{3p} electron pair, as discussed for HF in Section 7.2. Cl is considerably less electronegative than F, raising the Cl AO energy levels relative to H. The bonding is less polar and more covalent. The Cl orbitals are larger than those of F, and are more disparate vs H $1s$; this makes the MO's larger in the vicinity of Cl. Both the MO and VB models predict a bond order of 1, a σ bond, and three lone pairs. In both HF and HCl there is a small interaction between $1s$ and ns; the three σ orbitals are given by linear combinations of the form $\sigma = a1s_H + b3s_{Cl} + c3p\sigma_{Cl}$, with $b << a,c$ due to the energy mismatch.

21. OH is intermediate between NH and HF; its electron configuration is

$$(\sigma_{2s})^2(\sigma_{2p})^2(n_{2p})^3;$$

the bond order is still 1 as before, and there is now 1 unpaired electron, making OH paramagnetic.

23. An MO energy-level diagram and the MO boundary surfaces are given at the right. Note that the Li valence AO levels are much higher in E than H $1s$; from Mulliken's second rule it suffices to combine $1s_H$ and $2s_{Li}$, giving an electron configuration $(\sigma_{2s})^2$, a single bond, with no unpaired electrons, in agreement with Lewis / VB. The orbitals are lopsided as discussed in Section 7.2, the σ_{2s} LUMO favoring H and the σ_{2s}* HOMO favoring Li.

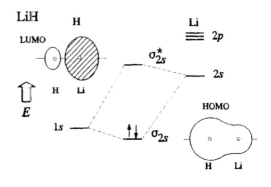

LiH is known to be a powerful reducing agent, implying that the HOMO is involved in its chemistry, as discussed near the end of Section 7.2. This fits with the loop-&-arrow mechanism for the reaction of LiH with water, Equation 5.28, in which the H end of LiH donates electrons to water. The LUMO of water must then also take part (See Section 7.3).

25. CH_2 has 6 valence electrons; using the level scheme given in Figure 7.17 we find that for linear CH_2 the configuration is $(\sigma_{2s})^2(\sigma_{2p})^2(n_{2p})^2$, with two unpaired electrons. Upon bending the configuration becomes $(a_1\sigma)^2(b_2)^2(a_1n)^2$; since both the $a_1\sigma$ and a_1n MO's drop in energy while only the b_2 rises, we predict that CH_2 will be a bent molecule with

no unpaired electrons. The HOMO orbital should resemble the a_1n orbital of Figure 7.17. Although this is the form methylene appears to take in organic reactions, in the gas phase the configuration $(a_1\sigma)^2(b_2)^2(a_1n)^1(b_1)^1$ is lower in energy by about 10 kcal/mol; the electrons prefer not to pair, owing to the small separation between the a_1n and b_1 levels. Even in this configuration the molecule is expected to be bent, since three electrons decline in energy upon bending, two increase, and one remains constant. The experimental bond angle is $136°$.

27. Linear H_3 is a simplified version of the XH_2 problem illustrated in Figure 7.16; the nodal rules governing the σ orbitals follow the Particle-in-a-Box. When X = H, only 1s AO's are available for building MO's, and we shall obtain 3 σ MO's from the 3 AO's with zero, one, and two nodes. These can be formed by taking linear combinations in the same way as indicated in Figure 7.16, except that in the one-node case the central AO must be omitted. The results are shown at the right.

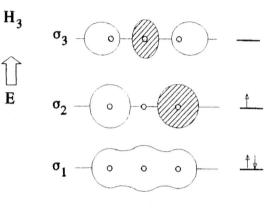

29. Resonance forms are shown at the right. In the MO picture, the migrating π bond is represented by a single delocalized π MO, also shown at the right. The O atoms become equivalent, making the π bond contribution to each N---O bond identical. For NO_2^- this means two identical bonds with a bond order of $1\frac{1}{2}$ for each; for NO_3^- $1\frac{1}{3}$, three identical bonds; thus we expect NO_2^- to have shorter, stronger bonds.

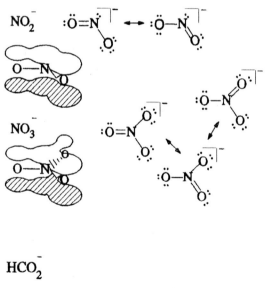

31. The resonance forms and delocalized π bond for HCO_2^- are shown at the right. Except for the C–H bond, the formate ion is isoelectronic with NO_2^-, which was discussed in Exercise 29 above. The σ bonding is described in VB theory by using sp^2 hybrid orbitals on C, and allowing them to overlap with 1s on H and $2p\sigma$ on O, for each of the two resonance forms. In VB theory the lone pairs must remain localized on a

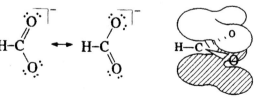

particular O atom, to allow a localized π bond to form. If the π bond is delocalized, the lone pairs must be as well, and therefore cannot be handled by VB.

8

Molecular Motion and Spectroscopy

Your Chapter 8 GOALS:

- Define a degree of freedom
- Identify the three types of molecular motion
- Calculate the number of the various degrees of freedom in a molecule
- Define and calculate the rotational analogues of linear momentum and kinetic energy
- Define moment of inertia
- Apply quantization concepts to rotational motion in order to understand the rotational constant, selection rules, and spectroscopy
- Use the harmonic oscillator model to describe molecular vibrations
- Manipulate energy and force constant expressions within the harmonic oscillator approximation
- Qualitatively describe vibrational spectroscopy
- Depict electronic transitions as occurring between two potential energy surfaces
- Relate photoelectron spectroscopy to orbital structure and orbital energy
- Learn the typical energy spacing and spectral region for rotational, vibrational, and electronic transitions
- Qualitatively explain the shape of the spectrum that results from $\sigma \rightarrow \sigma^*$ and $\pi \rightarrow \pi^*$ electronic transitions

Chapter 8 KEY EQUATIONS:

- 8.4, 8.6, 8.7, 8.9, 8.10, 8.13, 8.14, 8.23, 8.24, 8.25, 8.28, 8.29

Overview

This solutions manual is meant to give you the answers to certain Exercises. At times, though, it has been designed to provide comfort or support in your moments of frustration. Having taught this material for many years, your authors have heard from many students who are dismayed at the amount of new material that is presented in a college chemistry class. Nowhere is there more new material than in Chapter 8, which is the culmination of Chapters 1 through 7. (As you look ahead, at least Chapters 9, 10, 12, and 13 contain material with which many of you are familiar.) The material described in this chapter, however, consists of the very concepts that are used by experienced chemists who are during current research. In fact the device in Figure 8.11 resides in our research laboratory! You are, however, prepared to deal with much of it, armed as you are with the quantization idea. All three of the transition types, vibrational, rotational, and electronic, are subject to quantized motions and energies. You can master the equations with patience. The *really* new material in this chapter is the degree of freedom considerations, which we discuss below. The deep thinker might have wondered why translational transitions are not discussed in Chapter 8, or do they even occur? They do occur, but the spacing between the various translational energy levels is so close that the states essentially comprise a continuum. An analogous situation is that which exists in the Earth-Sun system. In Exercise 31 of Chapter 2 we found that $\Delta n = 1$ produced a negligible change in the system. The same argument applies to transitions between different translational energy levels. This does *not* mean that the translational *energy* is not important (see Chapter 9), nor does it mean that the *number* of translational states is not significant. Appendix D deals with the number of translational states, especially in Equations D.21, 22, 27, and 28. What it *does* mean is that is translational transitions are not observable in any ordinary spectroscopic way.

Concerns over degrees of freedom are not of quantum origin. Parceling energy into various degrees of freedom is a result of the classical kinetic molecular theory, developed by Maxwell in 1859 (see Chapter 9). This equal partitioning of energy (the *equipartition* theorem, Section 10.7) among the degrees of freedom was used to explain various physical properties of compounds, the Law of Dulong and Petit (Section 10.7) being one example. Table 8.1 will help you to solve Exercises 1, 2, and 3, which require you to determine the number and kinds of degrees of freedom that exist in molecules. In Chapter 8, however, we delve deeper than Maxwell could have to show that these degrees of freedom are, in fact, subject to quantum effects. Further, these quantum effects are observable in the spectroscopic study of molecules.

The discussion in Section 8.2 that runs up to the *Quantization of rotational energy* subsection is purely classical; the rotating molecule is held together by a rigid bond, creating a *rigid rotor*. When the atoms are arranged in such a manner, torques that act on the atoms create circular (or, angular) motion. From this analysis we derive classical expressions for the kinetic energy of rotation K_{rot}, the moment of inertia I, the all-important angular momentum J, and the rotational period τ_{rot}, which is another one of those benchmark numbers that you ought to know: $\tau_{rot} \approx 10^{-12}$ sec at room temperature. Exercises 4 through 7 deal with this subsection. Note that Exercise 6 can be solved via an engineering formula that is derived from Equation 8.9:

$$\tau_{rot} = 4.87245 \times 10^{-12} \sqrt{\frac{\mu(\text{amu}) \times r_e(\text{angstroms})}{T(\text{K})}}$$

We urge you to confirm this formula for yourself. For H_2 and I_2 the atoms in (classical) rotation would rotate in the exact manner as the ends of a baton about the center of mass, which would be where the twirler balances the baton. The HI motion would resemble a planet (the H atom) around the Sun (the I atom) since the center of mass nearly coincides with the I nucleus. In the remaining parts of Section 8.2, we "turn on" the quantum effects that are present in rotation. The disturbing result of this is immediately apparent in Exercise 8. In H_2 the deBroglie wavelength is more than half of $2\pi r_e$, which we regard as a confinement distance, and significant quantum effects are expected. Therefore, the baton model of which we just spoke is simply not true for (at least) light molecules or diatomic molecules involving H, and no classical picture can describe what is going on in these cases. Exercises 11 and 12 explore this topic further. To solve Exercise 12, substitute $\varepsilon_j = k_B T$ and the given approximation into Equation 8.11. Additionally, be sure that you grasp that the fact that the spacing between adjacent rotational levels is $2Bj$ (Equation 8.14 and Figure 8.3), but that a pure rotational spectrum, such as that of HCl in Figure 8.4, gives lines that are spaced by $2B$. In this model of rotational motion the B value that we obtain is crucial in determining bond lengths. See Example 8.1 and Exercises 13 and 14. In Exercise 14 you should find that $B = 6.59$ cm^{-1}.

The same classical to quantum approach is used in Section 8.3 for vibrational motion, and the classical description is explored in Exercises 16, 17, and 18. The widely used harmonic oscillator model that is developed was seen previously in Example 1.4, and it has good success when a molecule is not at a level of high vibrational excitation. Equation 8.24 is the vibrational analogue of Equation 8.9, and it can be solved to demonstrate that $\tau_{vib} \approx 10^{-14}$. Note that the classical turning point is where the kinetic energy is zero and the potential energy is a maximum. At this point the oscillating particle will *turn* around and head back the other way. Another new term in the vibrational discussion is "normal mode". The formal definition of a normal mode involves symmetry arguments that we are not prepared to discuss. A less formal definition is that a normal mode is a fundamental motion of a molecule. A molecule may have several normal modes (CO_2 has four, Figure 8.2) and these can be added in various ways to yield even more complex vibrations. Two brief analogies illustrate this point. There are seven basic colors, thanks to Mr. Roy G. Biv, and seven fundamental SI units. *Every* color and *every* unit is or is derived from one of these seven building blocks. Such it is that *every* vibration *is* a normal mode or *is a combination of* several normal modes (as long as the motion is truly harmonic.) This becomes more of an issue when the molecules grow in complexity. In Chapter 8, however, we limit the discussion to small molecules and the true meaning of normal modes is not crucial.

The quantization of vibration explains the spectra we see and the origin of zero point vibrational energy (Equation 8.26), which exists even at absolute zero. The concept of zero point energy impacts the classical amplitude of vibration that is calculated in Exercise 20, where it is shown that vibrating objects of low mass and short bond lengths have high zero point energy. To solve this Exercise, one can define from Equation 8.23 an amplitude $a = (r_+ - r_-)/2 = [2\varepsilon/k]^{1/2}$ and then substitute the zero point energy $(1/2)hc\bar{\nu}$ into this expression for a. Exercise 21 is, perhaps, the most important Exercise in the entire book. Yes! We just said that. The answer to this question contains all of the pieces of the puzzle that scientists struggled to answer for scores and scores: the meaning of T, k_B, quantization, degrees of freedom, the occupation of degrees of freedom, the impact of degrees of freedom on heat capacity, ... OK. We'll stop, but be sure to understand this Exercise. The key to solving Exercise 25 is to note the "identical bond potential energy curves" and divine the implication of this fact. See the solution in this manual.

Electronic spectroscopy continues the focus that MO theory places on bond order. When electronic transitions result in equal numbers of bonding and anti-bonding electrons, dissociation results. If the number of bonding electrons exceeds the number of anti-bonding electrons, the molecule will stay together, but changes in bond strength and bond order result. (You should review Exercises 9, 10, and 11 of Chapter 7 if the reason for this is unclear.) Exercise 28 looks at this issue. Further, bound electronic excited states exhibit spectra with many lines because of excitation to various vibrational and rotational levels in the excited electronic state. Note, however, that Figure 8.10 shows that rotational transitions are impossible to see at low resolution because rotational transitions are energetically very closely spaced, as shown Figure 8.8. Equation 8.30 is used to solve Exercises 30 and 31 with the He photon energy of 21.22 eV and $-IE \approx E_{orbital}$. In particular the estimated photoelectron kinetic energies for Exercise 30 are 2.5, 4.5, and 5.6 eV.

1. This and the next two Exercises apply the formulas of Table 8.1, derived from the discussion of Section 8.1. To get the correct number of rotational and vibrational degrees of freedom (DFs) for molecules triatomic or larger, you must be able to deduce the geometry of the molecule—whether it is linear or bent—based on the structural principles of Chapter 6. For example, if SO_2 were linear, it would have one more vibration and one less rotation. See the text key for answers.

3. In "clusters" like $(H_2O)_n$ and $H^+(H_2O)_n$ the entire assembly acts as a single molecule; it is desirable to derive a general formula valid for any number (n) of monomer units. For $(H_2O)_n$, we have $T=3$, $R=3$, and $V=3(3n)-6 = 9n-6$; for $H^+(H_2O)_n$, again $T=3$, $R=3$, but $V=3(3n+1)-6 = 9n-3$. See the discussion for the two previous Exercises above and the text answer key.

5. This exercise is simply a substitution and algebraic simplification. Using Equation 8.2 in Equation 8.3, second line, we have

$$K_{rot} = \tfrac{1}{2} m_A \left(\frac{m_B}{m_A + m_B} r_e \right)^2 \omega^2 + \tfrac{1}{2} m_B \left(\frac{m_A}{m_A + m_B} r_e \right)^2 \omega^2$$

$$= \tfrac{1}{2} \frac{m_A m_B}{(m_A + m_B)^2} (m_A + m_B) r_e^2 \omega^2$$

$$= \tfrac{1}{2} \frac{m_A m_B}{m_A + m_B} r_e^2 \omega^2$$

In the second line all common factors have been taken out, leaving only $(m_A + m_B)$ to represent the two terms. After cancelling this factor, we may recognize the reduced mass $\mu = m_A m_B/(m_A + m_B)$ (Equation 8.5) and the moment of inertia $I = \mu r_e^2$ (Equation 8.6) to gain the canonical form $K_{rot} = \tfrac{1}{2} I \omega^2$. Note that in finding 8.4 from 8.2 and 8.3, we use the same algebra as would be required to demonstrate the double equality Equation 8.6. (Using $J=I\omega$ we may also obtain Equation 8.7, $K=J^2/(2I)$.)

7. The moment of inertia of CO_2 contains no contribution from C; in fact it is the same as that of an O_2 molecule with $r_e = 2(1.16) = 2.32$ Å. The moment of inertia is

$$I = 2m_O(r_{C=O})^2 = 2(16.00 \text{ amu})(1.16 \text{ Å})^2 = 43.1 \text{ amu Å}^2 = 7.15 \times 10^{-46} \text{ kg m}^2,$$

We then have

$$\tau_{rot} = 2\pi \left[\frac{7.15 \times 10^{-46} \text{ kg m}^2}{2(1.381 \times 10^{-23} \text{ J/K}) T(K)} \right]^{\frac{1}{2}}$$

$$= \frac{3.20 \times 10^{-11}}{\sqrt{T}}$$

This yields $\tau_{rot} = 1.85$ ps @ 300K, 0.584 ps @ 3000K, and 5.84 ps @ 30K.

9. In the Bohr model, the angular momentum, denoted L rather than J, is quantized

according to $L = nh/(2\pi)$ (see Exercise 2.29). In each Bohr orbit the radius is r_n; for an infinitely massive nucleus, as first assumed by Bohr, the moment of inertia is $I = m_e r_n^2$. Therefore, according to Equation 8.7, we have

$$K_{rot} = \frac{J^2}{2I} = \left(\frac{n^2 h^2}{4\pi^2}\right)\left(\frac{1}{2m_e r_n^2}\right)$$

$$= \frac{n^2 h^2}{8\pi^2 m_e r_n^2}, \quad n=1, 2, 3,\ldots$$

The electron mass can readily be replaced with the reduced mass μ of the electron-proton pair (see Equation 2.8) for greater accuracy. With μ in place of m_e the formula becomes analogous to the rigid-rotor Equation 8.13, where instead of n^2 we have $j(j+1) = j^2+j$, $j = 0,1,2,\ldots$ The Bohr formula for the rotational energy is replaced by a still more accurate formula due to Schrödinger (see Exercise 3.25) that is identical in form to Equation 8.13, with j replaced by l, the atomic angular momentum quantum number. This is so because the Schrödinger equation for the angular motion of the H atom is of the same form as that for the rigid rotor, since the Coulomb potential is radial, independent of angles.

11. The energy-level-spacing formula Equation 8.14, when multiplied by c and using the definition of B given below Equation 8.13, yields the Bohr transition frequency

$$\nu = \frac{\Delta\varepsilon_{j,j-1}}{h} = 2cBj = \frac{hj}{4\pi^2 I}$$

According to the classical theory of Newton and Maxwell, absorption will occur when the frequency of the light matches the frequency of oscillation of the molecular dipole, in this case the rotational frequency. According to Bohr's Correspondence Principle, this classical condition is attained for large values of the quantum number(s) which change during a transition. To find the rotational frequency as a function of j, we combine Equations 8.4, 8.8, and 8.11:

$$\nu_{rot} = \frac{1}{\tau_{rot}} = \frac{\omega_{rot}}{2\pi} = \frac{1}{2\pi}\left[\frac{2K_{rot}}{I}\right]^{\frac{1}{2}} = \frac{1}{2\pi}\left[\frac{2j(j+1)h^2}{8\pi^2 I^2}\right]^{\frac{1}{2}}$$

$$= \frac{h}{4\pi^2 I}\sqrt{j(j+1)}$$

Now as j becomes large, $j(j+1) \rightarrow j^2$, and this formula reduces to that for ν above. Note that the limit of large j is equivalent to the limit $\varepsilon_j \gg \Delta\varepsilon_{j,j-1}$, since ε_j goes as j^2 while $\Delta\varepsilon$ goes as j. In this same limit, where j is approximately a continuous variable, $\Delta\varepsilon$ becomes indistinguishable from $d\varepsilon/dj = B(2j+1)$.

13. See Example 8.1 and the text answer key.

15. The far-infrared or microwave regions of the spectrum correspond to rotational lines; the condition for the existence of a pure rotational spectrum is that the electric dipole moment $\mu \neq 0$. This in turn depends on molecular symmetry and geometry, as discussed in Chapter 6. ClF is asymmetric, a polar covalent bond, $\mu \neq 0$, and hence

will show a pure rotational spectrum. Cl_2 is symmetric, $\mu = 0$, and no spectrum is present. NO_2 is bent, $\mu \neq 0$, spectrum. CS_2 is linear and therefore symmetric, $\mu = 0$, no spectrum. C_2H_4 is symmetric about the midpoint of the $C\!=\!C$ bond, $\mu = 0$, no spectrum. CH_3Cl is asymmetric due to the $C\!-\!Cl$ bond, $\mu \neq 0$, spectrum. All these molecules will show rotational Raman spectra, however.

17. The Morse function, written so that $V(r_e) = -D_e$, is

$$V(r) = D_e[1 - \exp(-\beta(r - r_e))]^2 - D_e.$$

Using the Maclaurin series for $e^x \equiv \exp(x)$ with $x = -\beta(r - r_e)$ yields

$$V(r) = D_e[1 - (1 - \beta(r - r_e) + \tfrac{1}{2}\beta^2(r - r_e)^2 + ...)]^2 - D_e$$
$$= D_e[\beta(r - r_e) - \tfrac{1}{2}\beta^2(r - r_e)^2 + ...]^2 - D_e$$
$$= D_e\beta^2(r - r_e)^2 + ... - D_e$$

Provided cubic and higher terms are omitted, this form agrees with that of Equation 8.20 so long as we identify $D_e\beta^2$ with $\tfrac{1}{2}k$, or $k = 2D_e\beta^2$. If you later take physical chemistry, you may learn more about the Morse function.

19. From Equation 8.23 we find that $\Delta r = r_+ - r_- = 2[2\varepsilon/k]^{1/2}$. The Uncertainty Principle, taken as an equality—which works best in the ground state of any system—yields $\Delta p_r = h/(4\pi\Delta r)$ and, approximately,

$$K = \frac{(\Delta p_r)^2}{2\mu} = \frac{h^2}{32\pi^2\mu(\Delta r)^2} = \frac{h^2 k}{256\pi^2\mu\varepsilon}$$

where we have substituted successively for Δp_r and Δr. Since for the harmonic oscillator $\varepsilon = 2K$, we find

$$\frac{\varepsilon}{2} = \frac{h^2 k}{256\pi^2\mu\varepsilon} \quad \text{or} \quad \varepsilon^2 = \frac{h^2}{128\pi^2}\frac{k}{\mu} = \frac{h^2}{128\pi^2}(4\pi^2\nu^2) = \frac{h^2\nu^2}{32}$$

Taking the square root of both sides leads to $\varepsilon_0 \approx h\nu/(4\sqrt{2})$. This is smaller than the true zero-point energy $h\nu/2$, implying that Δr has been overestimated, but nonetheless this estimate of ε_0 shows how the energy-level dependence on $h\nu$ is established.

21. In cm^{-1} units we need the quantity $k_B T/(hc)$; $k_B/(hc)$ is the reciprocal of the second radiation constant hc/k_B, with a value

$$k_B/(hc) = (1.381 \times 10^{-23}\ J/K)\ /\ [(6.626 \times 10^{-34}\ J\ s)(2.998 \times 10^8\ m/s)(100\ cm/m)]$$
$$= (1.439\ cm\ K)^{-1} = 0.695\ (0.695035)\ cm^{-1}/K.$$

At 300 K, $k_B T/(hc) = (0.695\ cm^{-1}/K)(300\ K) = 209\ cm^{-1}$. This is smaller than most of the $\bar{\nu}$ values in Table 8.3. Since $\bar{\nu}$ measures the energy level spacing in cm^{-1}, the quantum of energy required to excite molecular vibration, $k_B T$ at 300 K does not suffice for most molecules, and they will be (predominantly) in their ground states. A small fraction will nonetheless exist in excited states, as dictated by the Boltzmann distribution, to be introduced in Chapter 10.

23. Vibrational transitions require a *changing* dipole moment, which allows symmetrical molecules such as CS_2 and C_2H_4 to join the list of the vibrationally active by way of vibrational modes that distort their symmetry: antisymmetric stretches and bends. The C−H bonds in C_2H_4, while classified by our criterion as pure covalent, do have a slight polarity $C^{-\delta}$—$H^{+\delta}$ which is enhanced by the π bond, and when vibrating asymmetrically these bond dipoles yield a net temporary dipole. (Such dipoles are of a different type than the *instantaneous* dipoles that give rise to nonbonded intermolecular attraction; the latter will be discussed in Chapter 14.) This leaves only Cl_2, which neither has nor can develop a dipole when vibrating, as being unable to absorb infrared radiation. Cl_2 does display a Raman vibrational spectrum.

25. Identical potential energy curves for the isotopes of a molecule imply identical "classical" bond dissociation energies D_e, where $V(r_e) = -D_e$, and force constants $k = d^2V/dr^2|_{r=r_e}$; hence only the reduced mass alters the frequency of vibration or the bond energy. Since $\nu \propto \mu^{-\frac{1}{2}}$, the ratio of frequencies of two isotopes is in the inverse ratio of the square roots of their reduced masses. (We shall see such a relationship again in Chapter 9, in Graham's Law.) From the isotope masses of Table 1.3, we find $\mu(^1H^{35}Cl) = 0.980$ amu, $\mu(^2H^{35}Cl) = 1.904$ amu, $\mu(^1H^{37}Cl) = 0.981$ amu; and $\mu(^{14}N_2) = 7.002$ amu, $\mu(^{14}N^{15}N) = 7.242$ amu. Now we may find, for example, $\bar\nu(^2H^{35}Cl)$ from the tabulated value in Table 8.3, which refers to $\bar\nu(^1H^{35}Cl)$:

$$\frac{\bar\nu_2}{\bar\nu_1} = \left[\frac{\mu_1}{\mu_2}\right]^{\frac{1}{2}} \quad \text{or} \quad \bar\nu_2 = (2990\,\text{cm}^{-1})\sqrt{0.980/1.904} = 2145\,\text{cm}^{-1}$$

The remaining values, given in the text answer key, may be found in a similar fashion. To find the corresponding bond dissociation energies, we use the vibrational zero-point energy relationship $D_e = D_0 + \frac{1}{2}h\nu = D_0 + \frac{1}{2}hc\bar\nu$. This relation must hold for each isotopic variant, so that, for example,

$$D_0(^2H^{35}Cl) = D_0(^1H^{35}Cl) + \frac{1}{2}hc(\bar\nu_1 - \bar\nu_2)$$
$$= 4.49\,\text{eV} + (6.20 \times 10^{-5}\,\text{eV/cm}^{-1})(2990 - 2145)\text{cm}^{-1} = 4.54\,\text{eV}$$

D_0s for the other isotopes are obtained similarly, and are given in the text answer key.

27. The valence electron configuration of HBr may be written as $(\sigma_{4s})^2(\sigma_{4p})^2(n_{4p})^4(\sigma_{4p}*)^0$ (see Chapter 7, particularly Exercise 7.19) where we have included the lowest-lying empty orbital $\sigma_{4p}*$, the LUMO. In order to dissociate HBr, the bond order must be reduced to zero; this may be accomplished by exciting a bonding σ_{4p} electron to the antibonding $\sigma_{4p}*$, a $\sigma \rightarrow \sigma*$ transition producing an excited configuration $...(\sigma_{4p})^1(n_{4p})^4(\sigma_{4p}*)^1$. The energetic situation is illustrated in Figure 8.10(a). The kinetic energy of the atomic fragments may be calculated using a variant of the photoelectric equation Equation 2.5, where the work function W is replaced by the bond energy D_0:

$$K = \frac{hc}{\lambda} - D_0 = \frac{1240\,\text{eV nm}}{266\,\text{nm}} - 3.751\,\text{eV} = 0.91\,\text{eV}.$$

29. Octatetraene has four double bonds; each of these contributes 2 π electrons, giving a total of 8. In the particle in a box model, the π energy levels are represented by those of a box with a length equal to the length of the molecular chain, given as 9.8 Å. The electrons are assume to occupy these levels in accord with the Pauli and Aufbau Principles, pairing in the lowest 4 box energy levels, as illustrated at the right. The level labeled π_5 is the lowest antibonding π orbital, and the transition must thus occur between $n=4$ and $n=5$ box energy levels. To apply Bohr's postulate $\Delta E = hc/\lambda$, we need to compute $\Delta E_{4\rightarrow 5}$. The particle in a box energy levels are given by (see Appendix B) $E_n = h^2 n^2/(8ma^2)$, so that

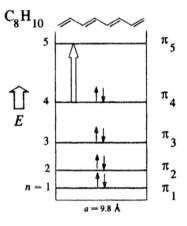

$$\Delta E_{4\rightarrow 5} = \frac{h^2}{8ma^2}(5^2 - 4^2) = \frac{9h^2}{8ma^2}$$

$$= \frac{9(6.626 \times 10^{-34}\,\text{J s})^2}{8(9.109 \times 10^{-31}\,\text{kg})(9.8 \times 10^{-10}\,\text{m})^2(1.602 \times 10^{-19}\,\text{J/eV})}$$

$$= 3.52\,\text{eV}$$

This energy falls into the near-ultraviolet region of the spectrum, with wavelength $\lambda = hc/\Delta E = (1240\,\text{eV nm})/(3.52\,\text{eV}) = 352\,\text{nm}$, 19% larger than the experimental value, 286 nm. The agreement is consistent with the relative crudeness of the model.

31. We use Equation 8.30 with $E_{\text{orbital}} \approx -\text{IE}$. The energy of the He photon is again $hc/\lambda = 1240\,\text{eV nm}\,/\,58.43\,\text{nm} = 21.22\,\text{eV}$. The IE, corresponding to the highest K_e, is IE $= hc/\lambda - K_e = 21.22 - 11.67 = 9.55\,\text{eV}$; the other energies are obtained similarly. The electron configuration of NO is $(\sigma_{2s})^2(\sigma_{2s}{}^*)^2(\sigma_{2p})^2(\pi_{2p})^4(\pi_{2p}{}^*)^1$; since $\pi_{2p}{}^*$ is the HOMO, it has the highest (least negative) $E_{\text{orbital}} = -9.55\,\text{eV}$. As K_e decreases, the orbital being ionized lies deeper and has a lower E_{orbital}. Since four bands appear in the spectrum of K_e, we can conclude that the $\sigma_{2s}{}^*$, σ_{2p}, π_{2p}, and $\pi_{2p}{}^*$ MOs are being ionized, and that σ_{2s} lies below the photoelectric threshold, $-E_{\text{orbital}} > 21.22\,\text{eV}$.

9

Properties of Gases
and the Kinetic Molecular Theory

<u>Your Chapter 9 GOALS</u>:

- Explain the origin of atmospheric pressure and the operation of a barometer
- Work with all pressure units: atmospheres, Pascals, torr (mm Hg)
- State and perform calculations with Boyle's Law, Charles's Law, and the ideal gas law
- Learn Dalton's Law of Partial Pressures and see the need for it in the kinetic molecular theory
- Use Graham's Law to relate mass and effusion
- Be able to sketch Maxwell-Boltzmann velocity distributions
- Recognize the Gaussian distribution
- Mathematically express the three velocities for gases that are defined in Chapter 9: root mean square, mean, and most probable
- Use collision rate theory to determine the mean free path of gas atoms or molecules
- Use the van der Waals equation to account for the non-ideal behavior of real gases
- Relate the van der Waals constants a and b to physical properties

<u>Chapter 9 KEY EQUATIONS</u>:

- 9.1, 9.4, 9.10, 9.13, 9.16, 9.19, 9.33, 9.36, 9.37, 9.39, 9.46, 9.47, 9.48, 9.55

<u>Overview</u>

In many ways Chapter 9 is a new beginning. Up until now we have mainly focused on the "microscopic" aspects of chemistry with Chapter 5 being a partial exception. Historically, however, chemistry's development began with macroscopic aspects. Recall, again, the quotation from Lord Kelvin at the start of Section 1.3. Science did not progress until it had numbers attached to its concepts, and in the early days of real science, macroscopic properties, such as pressure, volume, temperature, density, and mass, were usually what could be measured. (Fraunhofer did measure spectral lines, so we'll give him and other early spectroscopists "microscopic" credit.) As an example, it was not until, at least, fifty years (some chemists argue almost 100 years) after Maxwell derived Equation 9.46 that the gas velocities (a microscopic quantity) described by $f(v)$ were verified experimentally, yet the measurement of temperature, pressure, and volume of gases (macroscopic quantities) had been performed *well before* 1912! It was from these macroscopic measurements that the science of thermodynamics was born, as we'll see in Chapters 10, 11, and 12. We start with gases in Chapter 9, however, because Maxwell and Boltzmann used gases to develop and define the concepts of heat and temperature, and changes in these properties allow for the analysis of energy changes in chemical reactions.

It is very likely that you are quite familiar with Boyle's Law, Charles's Law, the ideal gas law, Graham's Law, Avogadro's Law, and the concept of density. For this reason, you should have an easy time with Exercises 8 through 20. Still, note in Exercise 18 that the identity of the 17 moles of gas that are formed does not matter, which is the whole point of the ideal gas law, Dalton's Law of Partial Pressures, and the kinetic molecular theory, and the calculated moles that form can be simply substituted into the ideal gas law to obtain T. One of the reasons that the gas laws can be presented fully in high school courses is that they are mathematically simple. This simplicity is due to equations such as Equation 9.13 being "equations of state", depending only on the current *state* of the system. For example, the temperature of the gas sample depends in no way on how it was heated or cooled to that temperature. The same notion applies to the amount of gas sample or its volume or pressure. The idea of equations of state (and state variables) is fundamentally important, and we will see much of this in Chapters 10 and 11.

Exercise 21 has a short solution, but it is a very important problem, illustrating the importance of Equation 9.21. For Exercise 22 assume that the water level inside the collection vessel is equal to the water level in the trough. With this being true, the pressure of the collected sample equals the atmospheric pressure. When the levels are not equal, a correction is required using the density of water, which is 1/13.6 the density of Hg. Therefore, a mercury column would be 1/13.6 as high as the water column. This column height must be subtracted (or added, as necessary) from (to) the atmospheric pressure. For Exercise 24 it is crucial to understand the other "useful corollary" of Dalton's Law of Partial Pressures: $P_i \propto n_i$ at constant T and V. For this reason, we may treat the partial pressures as if they were moles, multiplying them by the coefficients of the balanced equation in typical stoichiometry fashion. For Exercise 26, assume that ppm and % refer to fractions by volume rather than by mass and that they yield the mole fractions directly. Exercise 28 is challenging, but Equation 9.16 allows for the determination of the molecular mass of the hydrocarbon. Proceed from there by finding the moles of hydrocarbon from the combustion data.

In introductory courses you often find a plot such as Figure 9.12 and the "take home" message that molecular speeds increase as the temperature increases. Chapter 9 gives you a bit more of the truth by providing mathematical details and equations for the three velocities of note, which also indicate that at the same T more massive gases move more slowly. Equation 9.46 provides us the opportunity to revisit the concept of a distribution function, which we saw in Chapter 3 with distribution functions for the electron. It takes some thought, but the "cloud" of Figure 3.3(d) is *identical* in appearance for both the location of an electron in hydrogen and the particular velocities of a gas sample. The only difference is that the axes in Figure 3.3(d) are spatial coordinates (x, y, z), whereas the axes for the velocity distribution are the components of the velocities. This is the essence of Figure 9.6, which is a difficult figure for many to grasp. It may prove to be helpful if you compare the volumes of the spherical shells that are shown in Figure 9.6(b) and Figure 3.8: $4\pi v^2 dv$ and $4\pi r^2 dr$, respectively. These differ only in the variable. The mathematical construct is the same. Further, note how Equation 9.47 is derived; we find the maximum of $f(v)$ by setting df/dv equal to zero. This is mathematically *exactly* what we did in Exercise 17 of Chapter 3 to find the most likely points in space for a $2p_z$ electron. The only difference there is that the derivative was taken with respect to r, not v. Thus, the important science/math point is that distribution functions appear in many places in science, but the math and meaning behind them are the same in all cases. This topic is explored in Exercise 37, where we convert from a velocity distribution to an energy distribution.

To solve Exercise 34, use the fact that reaching escape velocity is equivalent to achieving $E = K + V(r) = 0$. Thus, $\frac{1}{2}mv^2 = GMm/r$, which can be solved for v. Use the definition of g, $g = GM/r^2$, to obtain the formula in the text answer key. Exercise 35 continues Exercise 34 and contains some mathematical manipulation that shows you how to work with very small numbers. This is a valuable skill to have by itself, but expressing numbers as what they really are, *not zero*, can be very illustrative. Even if you are not assigned this Exercise, you should master the process that is shown. Exercise 38 is made easy by the engineering formula that is developed in Exercise 33. The solution to Exercise 40 requires Equations 9.49, 9.50, and 9.51. All of these contain the number density [X], which for an ideal gas can be expressed as P/k_BT. The derivation for this uses Equation 9.13, and you should perform it yourself. Further, note that the velocity to use for Exercise 40 is from Equation 9.48, one of three presented in the Chapter. Be sure that you know the difference between these velocities. Exercises 41 and 42 use the same concepts and equations that are needed in Exercise 40. In particular Exercise 42 should be started by converting the mass loss in a given time into an atomic effusion rate. This should then be used to solve Equation 9.49 for the number density of K. Exercises 43 and 44 are Graham's Law problems, and M(air) was found in Exercise 7.

The comparison that is sought in Exercises 47 and 48 relies primarily on the difference in the a coefficients between N_2 and He. As indicated in Example 9.5, the repulsive correction to non-ideality is dominant only for H_2, He, and Ne, which is the reason that these gases are the only ones in Table 9.3 with a coefficients less than one. For this reason you will find that the ideal gas equation yields a pressure that is *higher* than the actual pressure for N_2 and *lower* than the actual pressure for He. (Loosely stated, N_2 molecules get along better with each other than He atoms do!) This concept of non-ideal behavior is probably foreign to you now, but in Section 12.7 and Exercises 75 through 78 of Chapter 12 it will be seen again in the guise of fugacity, which corrects for non-ideal pressures, and ionic strength, which deals with non-ideal concentrations. Advanced courses also pursue this further.

As you perform the iterations of Exercise 50, you should find that the molar volumes progress from 1.6411 to 1.5239 and so on until convergence is reached. If these are not your first values, you have not done the calculation right. (Convergence is a fancy term that means the value which you are calculating approaches a constant value or does not change with continued iterations. The concept of convergence is crucial to the theorems of integral calculus.)

9 Properties of Gases and the Kinetic Molecular Theory

1. Solving Equation 9.4 for h gives $h = P/\rho g$. Thus,

$$h = \frac{101,325\ \text{Pa}}{(1.00\ \text{g/cm}^3)(10^{-3}\ \text{kg/g})(100\ \text{cm/m})^3(9.807\ \text{m/s}^2)} = 10.3\ \text{m}\ (34\ \text{ft})$$

This result is well-known in the mining industry; a suction pump cannot draw water out of a mineshaft more than 34 feet below ground.

3. From Equation 9.4,

$$eh = \frac{P}{\rho g} = \frac{101,325\ \text{Pa}}{(1.29\ \text{kg/m}^3)(9.807\ \text{m/s}^2)} = 8010\ \text{m}\ (8.01\ \text{km})$$

5. If $P = \rho g h$, then $\Delta P = \rho g \Delta h$, where Δ represents a difference. Furthermore, we are looking for the relative Δh's for oil and Hg at the same ΔP. Thus

$$\Delta h_{\text{Hg}} = \Delta h_{\text{oil}}\left(\frac{\rho_{\text{oil}}}{\rho_{\text{Hg}}}\right) = (150.\ \text{mm})\left(\frac{1.09\ \text{g/cm}^3}{13.595\ \text{g/cm}^3}\right) = 12.0\ \text{mm Hg} \equiv 12.0\ \text{torr} = \Delta P$$

$$\Delta P(\text{atm}) = 12.0\ \text{torr}(1\ \text{atm}/760\ \text{torr}) = 0.0158\ \text{atm}$$

7. Equation 9.16 gives $\rho = MP/RT$. Since the molar mass $M = N_A m$, where m is the molecular mass, and $R = N_A k_B$, where k_B is Boltzmann's constant, $\rho = mP/k_B T$ as well. In the following, we substitute ρ in the differential relation $dP = -g\rho dz$, separate variables, and integrate at constant temperature:

$$dP = -g\left(\frac{mP}{k_B T}\right)dz$$

$$\frac{dP}{P} = -\frac{mg}{k_B T}dz$$

$$\int_{P_0}^{P}\frac{dP'}{P'} = -\frac{mg}{k_B T}\int_0^h dz$$

$$\ln P'\ |_{P_0}^{P} = -\frac{mg}{k_B T}z\ |_0^h$$

$$\ln\left(\frac{P}{P_0}\right) = -\frac{mgh}{k_B T} \quad\text{or}\quad P = P_0 e^{-mgh/k_B T}$$

The logarithmic integral $\int dx/x = \ln x + C$ appears frequently in science. Note that mgh is the gravitational potential energy of a single molecule; the exponential is an example of a "Boltzmann factor" (see Chapter 10 and Appendix D). A similar relation holds for the molar concentration $c = n/V = P/RT$ and density ρ as functions of altitude. For the calculation we need the mass m; we use the average mass of air calculated from Table 9.2 for the three major constituents N_2, O_2, and Ar as

$$M_{\text{air}} = 0.781(28.01) + 0.209(32.00) + 0.0093(39.95) = 28.9\ \text{g/mol}$$

We invoke the above relations between m, k_B and M, R to rewrite $mgh/k_B T$ as Mgh/RT, where M is expressed in kg/mol, allowing the calculation

$$P = (1\ \text{atm})\exp\left[-\frac{(28.9\times 10^{-3}\ \text{kg/mol})(9.807\ \text{m/s}^2)(8\times 10^3\ \text{m})}{(8.314\ \text{J/K mol})(298.15\ \text{K})}\right] = 0.401\ \text{atm}$$

At this pressure, the reduced P_{O_2} makes breathing difficult for all save the hardiest mountain climber. This treatment is approximate; its most glaring shortcoming is the neglect of the temperature profile of the atmosphere. For $h < 12$ km (in the *troposphere*), the temperature drops roughly linearly with altitude, averaging about 240 K at 8 km. This can be accounted for by inserting a linear $T(z)$ in the above RHS integral, instead of pulling T outside it; the result is that P drops more steeply with altitude. At higher altitudes, the atmosphere begins to *fractionate*, with lighter gases rising to the top, while solar radiation atomizes O_2, giving rise to a new chemistry in the *stratosphere* ($h = 12$ to 50 km).

9. To avoid decimals, we take atmospheric pressure to be 15 lb in^{-2} (15 psi). To get absolute pressures, we therefore add 15 to the pressures in psig. Thus the volume of air within the compressor is given by Boyle's Law as

$$V_{compressor} = \frac{(47\,\text{psi})(160\,\text{L})}{(135\,\text{psi})} = 56\,\text{L}$$

and the volume occupied at 1.0 atm (15 psi) is

$$V_{atm} = \frac{(47\,\text{psi})(160\,\text{L})}{15\,\text{psi}} = 500\,\text{L}$$

The last calculation could also have used the compressor pressure and its computed volume. Note that unit systems can be mixed, since only ratios are involved.

11. The simple answer, which requires no computation, is that, after cooling, the pressure inside the thermos has dropped far enough below atmospheric pressure as to make it impossible to open. To be more quantitative, note that the volume is constant. Equation 9.10 then gives $P_1/T_1 = P_2/T_2$. Now, assuming the pressure was 1.00 atm when the thermos was filled, the new pressure for the colder gas is $P = (1.00\,\text{atm})(295\,\text{K})/363\,\text{K} = 0.81$ atm. The net inward force on the cap is given by

$$F_{net} = [(1.00 - 0.81)\,\text{atm}](101{,}325\,\text{Pa / atm})(30 \times 10^{-4}\,\text{m}^2) = 57\,\text{N}\,(13\,\text{lb})$$

In this scenario, a large role is also played by the vapor pressure of the tea water, which decreases greatly on cooling, and creates an even better partial vacuum in the cooled thermos.

13. The trapped volume at point A is

$$V = \pi r^2 h = \pi(0.375/2\,\text{in})^2(48/4\,\text{in})(2.54\,\text{cm/in})^3 = 21.7\,\text{cm}^3 = 0.0217\,\text{L}.$$

Then from the ideal gas law Equation 9.13,

$$n_{air} = \frac{PV}{RT} = \frac{(29.125\,in\,Hg)(25.4\,mm/in)(1\,atm/760\,mm\,Hg)(0.0217\,L)}{(0.08206\,L\,atm/K\,mol)(288\,K)}$$

$$= 8.94 \times 10^{-4}\,mol = 0.894\,mmol$$

$$m_{air} = nM = (0.894\,mmol)(28.9\,mg/mmol) = 25.8\,mg$$

The molecular mass of air was taken from the calculation of Exercise 7.

15. From Equation 9.13, carrying the maximum number of significant figures our knowledge of R allows,

$$V_{NTP} = \frac{(1\,mol)(0.0820574\,L\,atm/K\,mol)(298.15\,K)}{1\,atm} = 24.465_4\,L$$

17. The balanced equation is $2Mg(s) + O_2(g) \rightarrow 2MgO(s)$. Equation 9.13 gives

$$n_{O_2} = \frac{PV}{RT} = \frac{(2.3\,atm)(2.6\,mL)}{(0.08206\,L\,atm/K\,mol)(299\,K)} = 0.244\,mmol$$

The mass of Mg is then

$$m_{Mg} = (0.244\,mmol\,O_2)\left(\frac{2\,mmol\,Mg}{1\,mmol\,O_2}\right)(24.305\,mg/mmol) = 11.8\,mg$$

19. To get a molar mass, one needs the mass and moles of the sample. The mass is given. The number of moles of vapor that condensed to yield the 2.428 g of liquid is determined from Equation 9.13,

$$n = \frac{PV}{RT} = \frac{(753\,torr)(1\,atm/760\,torr)(0.400\,L)}{(0.08206\,L\,atm/K\,mol)(373\,K)} = 0.01295\,mol$$

Thus the molar mass is $M = 2.428\,g/0.01295\,mol = 188\,g/mol$.

21. The mole fraction of O_2 from Table 9.2 is 0.20942. Using Dalton's Law, written in the form of Equation 9.21, we find

$$P_{O_2} = X_{O_2}P \quad or \quad P = \frac{P_{O_2}}{X_{O_2}} = \frac{100\,torr}{0.20942} = 478\,torr$$

23. Dalton advises us to treat each gas separately; at constant T we may use Boyle's Law for the expansion of each component into the other's volume. The total volume after mixing $V_2 = V_{1,He} + V_{1,Ar} = 5.0 + 3.0 = 8.0\,L$ is the same for each component. Thus

$$P_{2,He} = \frac{P_{1,He}V_{1,He}}{V_2} = \frac{(210\,torr)(5.0\,L)}{8.0\,L} = 131\,torr$$

$$P_{2,Ar} = \frac{P_{1,Ar}V_{1,Ar}}{V_2} = \frac{(320\,torr)(3.0\,L)}{8.0\,L} = 120\,torr$$

The total pressure $P_2 = P_{2,He} + P_{2,Ar} = 131 + 120 = 251\,torr$, and the mole fractions are

$$X_{He} = \frac{P_{2,He}}{P_2} = \frac{131 \text{ torr}}{251 \text{ torr}} = 0.52 \quad \text{and} \quad X_{Ar} = 1 - X_{He} = 0.48$$

25. This exercise sets the stage for the combustion of Exercise 18. The balanced gas-phase combustion equation is $2C_8H_{18} + 25O_2 \rightarrow 16CO_2 + 18H_2O$. For this stoichiometry we find

$$n_{O_2} = 0.025 \text{ mL } C_8H_{18} \left(\frac{0.74 \text{ g}}{\text{mL}}\right)\left(\frac{1 \text{ mol}}{114.2 \text{ g}}\right)\left(\frac{25 \text{ mol } O_2}{2 \text{ mol } C_8H_{18}}\right) = 2.02 \times 10^{-3} \text{ mol}$$

Equation 9.13 (see Exercises 13-19 for its use) with $T = 323$ K, $V = 0.050$ L, and $n_{O_2} = 2.02 \times 10^{-3}$ mol yields $P_{O_2} = 1.073$ atm. The stoichiometric partial pressure of gasoline vapor is $(2/25)(1.073 \text{ atm}) = 0.086$ atm. If O_2 is present at its normal mole fraction in air, then $P_{air} = P_{O_2}/X_{O_2} = 1.073 \text{ atm} / 0.20942 = 5.125$ atm (see Exercise 21), and $P = P_{air} + P_{C8H18} = 5.125 + 0.086 = 5.21$ atm. These would be the conditions in a cylinder under compression but before firing the sparkplug. The partial pressures of air and gasoline found here lead to the often-quoted 15:1 mass ratio of air to fuel for optimum efficiency.

27. In stoichiometric problems, the ideal gas law Equation 9.13 becomes a tool for finding moles of gas, $n = PV/RT$. In Chapter 1 you already used this tool under STP conditions; now the same calculations can be made for any T and P. Here the balanced chemical equation is $Fe(s) + H_2SO_4(aq) \rightarrow FeSO_4(aq) + H_2(g)$. The balloon volume is $V = \frac{4}{3}\pi r^3$. With the aid of Equation 9.13 we find

$$n_{H_2} = \frac{(0.960 \text{ atm})[\frac{4}{3}\pi(2.5 \text{ m})^3](10^3 \text{ L/m}^3)}{(0.08206 \text{ L atm/K mol})(285 \text{ K})} = 2690 \text{ mol}$$

$$m_{Fe} = 2690 \text{ mol } H_2\left(\frac{1 \text{ mol Fe}}{1 \text{ mol } H_2}\right)\left(\frac{55.845 \text{ g Fe}}{\text{mol Fe}}\right) = 1.50 \times 10^5 \text{ g (150. kg)}$$

29. The mass of Cr combined with the balanced equation gives us moles of Cl_2; then Equation 9.13 yields V:

$$n_{Cl_2} = 1 \text{ kg Cr}\left(\frac{1 \text{ mol Cr}}{52.00 \times 10^{-3} \text{ kg Cr}}\right)\left(\frac{1 \text{ mol } CrO_2Cl_2}{1 \text{ mol Cr}}\right)\left(\frac{7 \text{ mol } Cl_2}{4 \text{ mol } CrO_2Cl_2}\right) = 33.66 \text{ mol}$$

$$V_{Cl_2} = \frac{(33.66 \text{ mol})(0.08206 \text{ L atm/K mol})(323 \text{ K})}{1.20 \text{ atm}} = 743 \text{ L}$$

31. In the kinetic theory, force and pressure are proportional to the product of the momentum change upon striking the container wall $\Delta p = 2mv$ and the frequency of wall collisions $1/\Delta t = v/(2L)$ (Equations 9.23 and 9.24). At a fixed temperature, $v \propto m^{-\frac{1}{2}}$. Therefore $\Delta p \propto m^{\frac{1}{2}}$ and $1/\Delta t \propto m^{-\frac{1}{2}}$, and their product is constant, independent of m, implying the same pressure for each gas. Thus, although He atoms impart less momentum per wall collision, they move faster and strike the wall more often than the heavier Xe atoms.

33. To find average velocities $\bar{v}$ from Equation 9.48, either one can use Boltzmann's constant with the mass m of one molecule in kg, or one can exploit $R = k_B N_A$ and use a

mass M in kg / mol. We will show both methods, culminating in one of our infamous "engineering formulas":

$$\bar{v} = \left[\frac{8k_BT}{\pi m}\right]^{1/2} = \left[\frac{8N_Ak_BT}{\pi M}\right]^{1/2}$$

$$\bar{v}(\text{m/s}) = \left[\frac{8(6.02214 \times 10^{23}/\text{mol})(1.38065 \times 10^{-23}\,\text{J/K})T}{\pi M(\text{g/mol})(10^{-3}\,\text{kg/g})}\right]^{1/2}$$

$$= 145.508\sqrt{T/M(\text{g/mol})}$$

or

$$\bar{v} = \left[\frac{8RT}{\pi M}\right]^{1/2}$$

$$\bar{v}(\text{m/s}) = \left[\frac{8(8.31447\,\text{J/K mol})T}{\pi M(\text{g/mol})(10^{-3}\,\text{kg/g})}\right]^{1/2} = 145.508\sqrt{T/M(\text{g/mol})}$$

The second method eliminates a factor of N_A, but of course the engineering versions are identical. Using the engineering formula, we readily find

$$\bar{v}(\text{H}) = 145.5\sqrt{2800/1.008} = 7670\,\text{m/s} = 7.67 \times 10^5\,\text{cm/s}$$

$$\bar{v}(\text{I}_2) = 145.5\sqrt{387/2(126.9)} = 180.\,\text{m/s} = 1.80 \times 10^4\,\text{cm/s}$$

$$\bar{v}(\text{C}_8\text{H}_{18}) = 145.5\sqrt{473/114.2} = 296\,\text{m/s} = 2.96 \times 10^4\,\text{cm/s}$$

35. In finding the ratio $p = f(v_{esc})/f(\alpha)$, where $f(v)$ is the Maxwell-Boltzmann velocity distribution Equation 9.46, the normalization constants cancel. We use $v_{esc} = 11{,}178$ m/s from Exercise 34, and compute α from Equation 9.47:

$$\alpha = \left[\frac{2k_BT}{m}\right]^{1/2} = \left[\frac{2RT}{M}\right]^{1/2}$$

$$\alpha(\text{m/s}) = \left[\frac{2(8.31447\,\text{J/K mol})T}{M(\text{g/mol})(10^{-3}\,\text{kg/g})}\right]^{1/2} = 128.953\sqrt{T/M(\text{g/mol})}$$

$$= 128.953\sqrt{273.15/31.999} = 376.76\,\text{m/s}$$

This yields the square velocity ratio $(v_{esc}/\alpha)^2 = 880.235$. Then p is obtained from

$$p = (v_{esc}/\alpha)^2 \exp[-(v_{esc}/\alpha)^2 + (\alpha/\alpha)^2]$$

$$= 880.235e^{-879.235} = e^{\ln(880.235)-879.235} = e^{-872.455}$$

$$= 10^{-872.455/\ln(10)} = 10^{-378.902} = 10^{-379+0.098} = 1.25 \times 10^{-379}$$

where we have used $\alpha^2 = 2k_BT/m$ to simplify $f(v)$. If you try to use a calculator, or even a computer, to do this calculation, it will probably give you an error and say that the answer is zero, because this number is far too small to be handled. Observe carefully how it was done here, and never be handcuffed by your (literally) stupid calculator. There are only 2×10^{43} O_2 molecules in the atmosphere, implying that, if p is taken as a population ratio, not a single molecule can escape at STP. Even in the far reaches of the atmosphere, where the gas temperatures can be greater than 1000K, the probability of losing oxygen atoms into interplanetary space is still negligible.

37. In converting a probability distribution from one variable to another, all that is needed is

a functional relationship, an equation relating the variables. This is used both to supplant the old variable with the new one, and to relate the differentials, essential to maintaining probability distributions by establishing the relative densities of the two variables. In our case, a $v \to E$ transformation, where E is the kinetic energy of a molecule, we use $E = \frac{1}{2}mv^2$, or $v = \sqrt{2E/m}$. Based on $f(E)dE = f(v)dv$ and using $f(v)$ of Equation 9.46 we have

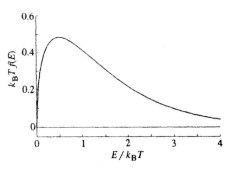

$$f(E) = f(v) \left| \frac{dv}{dE} \right|$$

$$= 4\pi \left(\frac{m}{2\pi k_B T} \right)^{3/2} v^2 e^{-\frac{1}{2}mv^2/k_B T} \left| \frac{dv}{dE} \right|$$

$$= 4\pi \left(\frac{m}{2\pi k_B T} \right)^{3/2} \frac{2E}{m} e^{-E/k_B T} \left| \frac{1}{2} \left(\frac{2}{m} \right)^{1/2} E^{-1/2} \right|$$

$$= \frac{2}{\sqrt{\pi}} (k_B T)^{-3/2} E^{1/2} e^{-E/k_B T}$$

A plot of this function in universal form is given above right. Note that the peak of this distribution occurs at $E = \frac{1}{2}k_B T$, its slope becomes infinite as $E \to 0$, and overall it is much less bell-shaped than the velocity distribution.

39. Using Equation 9.35 we have $E = \frac{3}{2}RT = (1.5)(8.314 \text{ J/K mol})(310\text{K}) = 3870$ J or 3.87 kJ/mol; dividing by 4.184 J/cal yields 0.924 kcal/mol. This energy accounts only for the translational degrees of freedom; in Chapter 10 we will learn about energy storage in CO_2 rotation and vibration.

41. Using Equation 9.49 for part (a) and 9.51 for part (b), we first calculate [air] and $\bar{v}_{air}$ for the conditions $P_A = 29.125$ in Hg $= 0.973$ atm at $T = 288$K:

$$[\text{air}] = \frac{N}{V} = \frac{P}{k_B T} = \frac{(0.973 \text{ atm})(101,325 \text{ Pa/atm})(10^{-2} \text{ m/cm})^3}{(1.3807 \times 10^{-23} \text{ J/K})(288 \text{ K})}$$

$$= 2.48 \times 10^{19} \text{ cm}^{-3}$$

$$\bar{v}_{air} = 145.5\sqrt{288/29.0} = 459 \text{ m/s} = 4.59 \times 10^4 \text{ cm/s}$$

a. At Point A:

$$Z_A = \frac{1}{4}[\text{air}]\bar{v}_{air}A$$

$$= \frac{1}{4}(2.48 \times 10^{19} \text{ cm}^{-3})(4.59 \times 10^4 \text{ cm/s})\{\pi[\frac{1}{2}(0.375 \text{ in})(2.54 \text{ cm/in})]^2\}$$

$$= 2.03 \times 10^{23} \text{ collisions s}^{-1}$$

We can use the proportionality between Z and P to find

$$Z_B = Z_A(P_B/P_A) = (2.03 \times 10^{23} \text{ s}^{-1})(111.5625/29.125) = 7.78 \times 10^{23} \text{ s}^{-1}$$

b. At Point A:

$$\lambda_A = (\sqrt{2}\,[air]\,\pi d^2)^{-1}$$
$$= [\sqrt{2}\,(2.48 \times 10^{19}\,cm^{-3})\,\pi(3.3 \times 10^{-8}\,cm)^2]^{-1}$$
$$= 8.3 \times 10^{-6}\,cm \quad (830\,\text{Å})$$

The inverse proportionality between λ and P leads to

$$\lambda_B = \lambda_A(P_A/P_B) = (830\,\text{Å})(29.125 \,/\, 111.5625) = 220\,\text{Å}$$

43. Adopting the general formula SO_x for the sulfur oxide, Graham's Law, Equation 9.39, implies that

$$\frac{t(N_2)}{t(SO_x)} = \left[\frac{M(N_2)}{M(SO_x)}\right]^{1/2} = \frac{34.3\,s}{51.8\,s} = 0.662$$
$$\therefore M(SO_x) = (28.0\,g/mol)/(0.662)^2 = 63.9\,g/mol$$

Since SO_x must contain at least one S atom, $M = 32.1\,g/mol$, the compound can only be SO_2.

45. Assuming a 1.00-m tube length, we may use Equation 9.39 to find the ratio of distances traveled in unit time, since distance is proportional to velocity. As the distances must also add up to 1.00 m, each can be determined. We have

$$\frac{x_{NH_3}}{x_{HCl}} = \left[\frac{M_{HCl}}{M_{NH_3}}\right]^{1/2} = \left[\frac{36.5}{17.0}\right]^{1/2} = 1.465$$
$$x_{NH_3} + x_{HCl} = 1.00\,m$$
$$2.465 x_{HCl} = 1.00\,m$$
$$x_{HCl} = 0.406\,m \quad \text{and} \quad x_{NH_3} = 0.594\,m$$

where in the third line we substituted for x_{NH3} from the Graham's-law ratio. This is in very good agreement with measurements using an evacuated tube, and still qualitatively correct for the experiment with an air-filled tube shown in the figure. If HI were substituted for HCl, the product NH_4I would form at a smaller distance from the acid source, since HI diffuses more slowly because of its greater mass; conversely, if ethylamine were substituted for NH_3, the deposit would be further from the HCl source because methylamine is more massive than NH_3. Quantitative estimates of the deposit distances follow the above sequence, with suitable mass replacements. The result for HI + NH_3 is $x_{HI} = 0.27$ m; and for HCl + $C_2H_5NH_2$, $x_{HCl} = 0.53$ m.

47. a. Equation 9.13 for the stated conditions easily gives $P = 159$ atm.
    b. See Example 9.5. Equation 9.55 with constants for N_2 from Table 9.3 yields

$$P = \frac{nRT}{V-nb} - a\left(\frac{n}{V}\right)^2$$
$$= \frac{(200.\,mol)(0.08206\,L\,atm/K\,mol)(291\,K)}{30.0\,L - (200.\,mol)(0.03860\,L/mol)} - (1.349\,atm\,L^2/mol^2)\left(\frac{200.\,mol}{30.0\,L}\right)^2$$
$$= 214.4 - 60.0 = 154\,atm$$

We note that, although the overall deviation from ideal behavior is only -3%, the breakout of repulsive and attractive terms shows a good deal of cancellation. That is, if a were zero, the deviation from ideality would be in the opposite direction and much larger ($+35\%$). Thus the attractive correction dominates, and the van der Waals (and presumably the observed) pressure is lower than ideal.

49. The calculations required to produce a graph are identical to those of the previous two Exercises, using a and b for NH_3. The graph is similar to that of Figure 9.3(a); in order to show more clearly the deviation from ideal behavior predicted by the van der Waals equation, we also include a plot like that of Figure 9.3(d) (shown as an inset in the P versus V graph). At the extreme Vs, we have

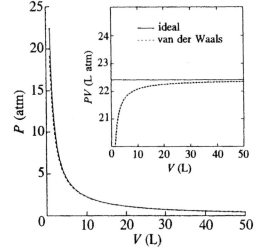

V(L)	P_{ideal}(atm)	P_{vdW}(atm)
1.0	22.4	19.1
50.0	0.448	0.447

The deviation is much larger at low V (high P); this is clarified by analyzing the two terms in the van der Waals expression for $P(V)$.

51. The Hg column is 111.56 inches ($P = 3.73$ atm) at point B; the ideal molar volume is then 6.34 L/mol, giving a PV product of 23.64 L atm/mol. Following the procedure of Exercise 47 for the van der Waals equation of state using a and b for N_2, we find the iterates 6.3203, 6.3200 L/mol, or a PV product of 23.57, an 0.3% difference. The data show about a 4.5% drop in PV at point B, much too large to be due to nonideal behavior.

53. The simplest approximation relating the van der Waals b to the molecular diameter d is $d \approx (b/N_A)^{1/3}$; for H_2O this yields

$$d = [(0.03048 \text{ L/mol})(1000 \text{ cm}^3\text{/L}) / (6.022 \times 10^{23} \text{ /mol})]^{1/3} = 3.70 \times 10^{-8} \text{ cm} = 3.70 \text{Å}$$

For H_2S, $d = 4.16$Å, and for H_2Se, $d = 4.30$Å. The diameter of H_2O from liquid density is 3.10Å (see Chapter 1). A more rigorous relationship between b and d is $b = \frac{2}{3}\pi N_A d^3$; this yields $d = 0.7816(b/N_A)^{1/3} = 2.89$Å for H_2O.

10

Energy Changes
in Chemical Reactions

Your Chapter 10 GOALS:

- State the Law of Conservation of Energy
- Define open, closed, and isolated chemical systems and the surroundings of a chemical reaction
- Define the pressure-volume work associated with gas producing reactions
- Relate heat and temperature change via the heat capacity of a system
- Explain what a state variable is
- State the First Law of Thermodynamics
- Define enthalpy
- Be able to use calorimetry to monitor the enthalpy change for a reaction
- Use standard heats of formation to calculate the enthalpy change of a reaction
- Use bond energies (enthalpies) to calculate the enthalpy change of a reaction
- Construct enthalpy level diagrams from heats of formation data or bond energy data
- Derive heat capacities from the number of degrees of freedom and the equipartition theorem
- Use the Boltzmann distribution to account for the very small contribution of vibrational degrees of freedom to the heat capacity

Chapter 10 KEY EQUATIONS:

- 10.6, 10.8, 10.10, 10.12, 10.16, 10.18, 10.20, 10.22, 10.23, 10.25, 10.27, 10.28, 10.30, 10.31, 10.32, 10.33, 10.35

Overview

The First Law of thermodynamics relates heat and work and thereby defines the internal energy of the system. In doing so, science has used *measurable quantities* like heat and work, to define the rather nebulous concept of "internal energy". Mass, density, inertia, angular momentum, and linear momentum, for example, are well defined. A precise definition of internal energy E, however, does not exist. Indeed, the First Law defines the *change* in the internal energy, without defining with any specificity what "internal energy" is. And so it is with entropy that we will meet in Chapter 11. Entropy, too, needs its own law, the Second Law of thermodynamics, and it is defined in terms of measurable quantities. Note that there are no "energy meters" or "entropy meters"; we only have thermometers, pressure gauges, graduated cylinders, etc..

Although enthalpy H is derived from E in typical presentations of thermodynamics in chemistry books, note that the First Law can be formulated in terms of E or H. Therefore, *Equations 10-18 and 10-23 are equivalent statements of the First Law*, where the difference in conditions, constant volume or constant pressure, accounts for the different subscripts on q. H, defined in Equation 10.22, is a very convenient, practical state variable because it is *defined* to be the heat transferred at constant pressure, a useful definition since this is the situation for systems that are open to the atmosphere on the top of a simple lab bench. In fact you have worked with this practical state variable many times in your previous chemistry courses. Assuming a constant external pressure, Exercises 16, 38, and 39 of Chapter 1 are enthalpy problems that are presented without mentioning the word. Exercises 13, 20, and 22 of Chapter 10 are examples of heat transfer processes at constant external pressure. Properties such as enthalpy are defined to help scientists characterize a system. There are other such properties, the Gibbs free energy, which we will meet in Chapter 11 and discuss for the rest of the book, and the Helmholtz free energy, which we won't get to cover, to name a few. Do not be mystified by these new properties which are mere contrivances that allow us to monitor a system by recording measurable quantities.

The concepts of work and heat are needed to solve the first 22 Exercises of the Chapter. Exercises 1, 2, and 3 require the use of Equation 10.8. Exercises 5 through 8 require the use of $q = mc\Delta T$, which is shown in Example 10.2. Exercise 6(b) gives you a chance to review your physics knowledge about units. If you need a refresher course, go to Chapter 13 in which volts and current are discussed at length. Exercises 10 through 13 are all based on the concept of heat flow from the hot body to the cold body described in each Exercise. That is, $q_{hot} = - q_{cold}$, followed by $q = mc\Delta T$. (This flow from hot to cold is, perhaps, met with a big 'of course'. We will return to this process in our introduction to Chapter 11.) For Exercise 10 you ought to be able to derive the engineering formula $T_f = [m_h T_{h,i} + m_c T_{c,i}]/(m_h + m_c)$. As you solve Exercise 13, don't forget to include the line over the C when you are indicating the *molar* heat capacity. It has been our experience that students ignore the importance of that symbol. Further, many of the Equations that appear in Section 10.3 can be rewritten in terms of molar heat capacities. Be sure that you understand the difference between $C_P = (5/2)nR$ and $\bar{C}_P = (5/2)R$. The same concept applies when you solve Exercises 20 and 22; you must divide by the moles of substance that are present to obtain the molar heat of reaction. Exercise 14(a) is an irreversible (see Chapter 11), isothermal expansion against a constant pressure, giving $\Delta E = 0$ from Equation 10.27. The First Law then gives $w = - q = - P_{ext}\Delta V$. Boyle's Law gives V_f. This solution, therefore, is what could be used for the expanding gas of the reaction shown in Equation 10.2. Exercise 14(b), however, is *entirely* different. We can't use Boyle's Law to get V_f because the process is not isothermal since $q = 0$. The key step is to assume that $P_{ext} = P_f$ and use the ideal gas law to express V_f, but this is a *much* harder problem.

Sections 10.5 and 10.6 examine two ways by which one can obtain the heats of various reactions. Both the standard heat of formation method, Equation 10.30, and the bond energy method, Equation 10.32, are based on Hess's Law. These calculations are generally simple, but you must be careful to use the coefficients of the balanced chemical equation. Solving Exercise 24 is an excellent way to test your understanding of not only how coefficients play a role in enthalpy calculations, but also how reversing equations affects the sign of ΔH. We urge you to take note that the enthalpy level diagram of this Exercise requires that an "extra" $O_2(g)$ be added to the highest and lowest energy levels to keep the equations balanced. *Many* students fail to understand this concept at exam time. Another example to consider is the comparison of the enthalpy level diagrams for Exercises 23 and 25. Nothing "extra" needs to be added for Exercise 23, but Exercise 25 requires that $NH_3(aq)$ be added to two levels. This addition is necessary to make the diagram correct, but it is even more crucial that you understand *why* it is needed. Experience has also taught your authors that students often do not understand the zero of enthalpy for both the standard heat of formation method and the bond energy method. We restate from the text the distinction. For the standard heat of formation, the zero of enthalpy is that point at which the reagents have been "transformed" into *elements in their most stable forms at NTP.* For the bond energy method, the zero of enthalpy is the point at which the reactants have been "transformed" into *free gas-phase atoms.* In both cases the reaction does not need to proceed by such a mechanism. Hess's Law, nonetheless, gives us the right ΔH since the overall enthalpy change can be represented by a sum of (sometimes imaginary) individual steps. Speaking of diagrams, Exercise 37 contains a whopper with a wealth of information. For Exercise 38 the heat of formation for $CH_4(g)$ is written as $C(s) + H_2(g) \rightarrow CH_4(g)$. This is one example of a formation reaction that you must master to be successful with the Chapter 10 material. In fact Exercise 42 tests your ability to write a formation reaction for 1,3-butadiene. Remember that it is *always* for the formation of 1 mol of product. Do whatever is required to the reactant coefficients to achieve this.

Exercises 43 through 51 return us to the degree of freedom concerns that were mentioned in the introduction of the Chapter 8 solutions, and here they are used to determine the heat capacities of gases. Now, classical physics did have it partially correct. Molecules do store energy in their various degrees of freedom, and those in which they can store energy are the translational, rotational, and vibrational degrees of freedom. For the record, we note that there are also electronic and nuclear degrees of freedom, but as in Chapter 8 we will not deal with them at this time, except to say that their contribution to the heat capacity is very small. Classical physics, however, did not recognize that the various degrees of freedom are subject to quantization restraints. What complicates the matter further is that the spacing is different for translational, rotational, and vibrational degrees of freedom. As discussed in Chapter 8 in the text and in the introduction in this manual, the spacing is smallest for translational energy levels and largest for vibrational levels. It turns out that at room temperature the spacing is small enough in translation that these quantum levels form a continuum. In line with the Bohr Correspondence Principle, the quantum numbers for the translational energy levels that are occupied at room temperature are on the order of 10^{10}. For this reason the kinetic molecular theory yields a *correct* "quantum number free" formula for the translational, that is, kinetic, energy of a monatomic gas, Equation 9.33 or 9.35. The rotational levels are farther apart, requiring a formula with quantum numbers, as in Equation 8.11. Likewise for vibration in Equation 8.25. Still, the rotational levels are close enough that thermal energy is sufficient to populate enough excited energy levels to allow for the classical predictions regarding heat capacity to be correct. This is not true for the vibrational levels. The N_2 example in Section 10.7 shows that thermal energy is *not* sufficient to excite even the first excited vibrational state, $v = 1$. (Compare this to the $n \approx 10^{10}$ state for translation.) Therefore, no energy can be stored in the vibrational degrees of freedom. The only energy that is present in vibration is the zero point energy. (From the above, do not think that an H_2 molecule does not vibrate at room temperature. It most certainly does – in its ground vibrational state.)

Figure 10.8 displays the effect that thermal energy has on the population of vibrational levels. Some chemists refer to k_BT as a "thermal quantum", indicating the role it plays in excitation to higher vibrational levels. Make no mistake, however. The "thermal quantum" is an analogy, and thermal energy does *not* come in bundles of k_BT. Note that the populations shown in Figure 10.8 actually require division of Equation 10.35 by the partition function q that is discussed in Appendix D. ($q \approx 1$ at 298 K, and $q = 1.77$ at 1000 K.) The qualitative significance of k_BT should still be evident, however. Calculation of q is explored in Exercise 51. When solving Exercise 46, remember that the molar heat capacity is $3R$ times the number of ionic particles in the formula. For Exercise 48 use Equation 9.13 to determine the number of moles of air, followed by Equation 10.10. Exercise 50 is a simple application of Equation 10.35.

We end by pointing out a fundamentally important distinction between the D_0 values of Table 8.3 and the bond energy values of Table 10.3, using the H_2 data. A comparison shows that Table 8.3 gives $D_0(H–H) = 103.3$ kcal/mol, whereas Table 10.3 gives $D_0(H–H) = 104$ kcal/mol, based on twice the atomization enthalpy. The difference lies in the fact that the Table 8.3 data is for $\Delta H°$ at 0 K, whereas the Table 10.4 data is for $\Delta H°_{298}$. This is a difficult distinction to grasp, but the extra thermal energy that is present at 298 K can be estimated using Equation 10.36, which is called Kirchhoff's Law by some chemists. Should you remain in chemistry, this will be encountered again in a physical chemistry class. (Laws for electric circuits also bear Kirchhoff's name. To avoid confusion, Equation 10.36 is not referred to as "Kirchhoff's Law" in *University Chemistry*.)

1. The "sudden" application of the compressive force is intended to convey a constant external pressure equal to the final pressure of the gas "during" the compression, $P_{ext} = P_2$. In that case, Equation 10.7 applies; we have

$$w = -P_2(V_2 - V_1) = -P_2V_2(1 - V_1/V_2) = -P_1V_1(1 - V_1/V_2)$$
$$= -(1.00\,atm)(6.00\,L)[1 - (6.00\,L)/(5.45\,L)](101.3\,Latm/J) = 61.3\,J$$

where we have used Boyle's Law to replace P_2V_2 by P_1V_1 (only possible at constant T). The positive value for w reflects work done *on* the system, always the case for compression. This is a *closed* system, with a trapped sample volume.

3. Here Equation 10.7 applies (as in Exercise 1), because the external pressure is constant, $P_{ext} = 745\,Torr = 0.980\,atm = P_2$. We have

$$w = -P_2(V_2 - V_1) = -P_2V_2(1 - V_1/V_2) = -P_1V_1(1 - P_2/P_1)$$
$$= -(170.\,atm)(50.0\,L)(1 - 0.980/170.)(101.3\,J/Latm) = -8.56 \times 10^5\,J$$

where we have used Boyle's Law twice in converting the work expression into one that uses only the information given. Incorporation of nonideal behavior (see Chapter 9) would involve altering the expanded volume V_2, which would have to be computed using n derived from the van der Waals equation of state for a given T as in Chapter 9. At 298K this procedure yields $w = -9.45 \times 10^5\,J$, about 10% more work, due to the repulsion between Ar atoms.

5. For this Exercise we measure the mechanical equivalent of heat in much the same way that Joule did by replacing his falling weight with a falling baseball! See Section 10.1. The maximum potential energy, $m_{ball}gh$, of the baseball has been completely converted to mechanical (kinetic) energy when it hits the catcher's glove. The heat energy, q, needed to increase the water temperature by $1\,°C$ is $m_{H2O}c$, where $c = 1.000\,cal/g\,°C$. Equating these energies for the case $m_{ball} = m_{H2O}$ leads to

$$gh = c \quad or \quad h = \frac{c}{g}$$
$$= \frac{(1.000\,cal/g\,°C)(1\,°C)}{9.807\,m/s^2}\left(\frac{4.184\,J}{1\,cal}\right)\left(\frac{1000\,g}{1\,kg}\right) = 427\,m$$

some 1400 ft. Inasmuch as the longest home run ever hit, reportedly by Babe Ruth in a spring training game, is 650 ft, it seems unlikely that such a high pop-up has ever been hit. We have also neglected air resistance. The Hall-of-Fame catcher Mickey Cochrane once caught a ball dropped from the top of the Washington Monument, 555 ft tall; the recoil broke his jaw, but the energy would have heated 140 g of water by only $0.4\,°C$.

7. If $q_w = q_{oct}$, then $m_wc_w\Delta T_w = m_{oct}c_{oct}\Delta T_{oct}$. With equal masses, we find

$$c_{oct} = c_w\frac{\Delta T_w}{\Delta T_{oct}} = (1.000\,cal/g\,°C)\frac{3.79\,°C}{9.63\,°C} = 0.394\,cal/g\,°C$$

9. The energy released in Exercise 3 was $8.56 \times 10^5\,J$. In Exercise 6 we found that 42.5 kcal were needed to heat enough water for two servings of tea. The number of servings

is then twice the ratio of these energies:

$$\# \text{ of servings } = \frac{8.56 \times 10^5 \text{ J}}{(42.5 \times 10^3 \text{ cal}/2 \text{ servings})(4.184 \text{ J}/\text{cal})} = 9.6 \text{ servings}$$

11. Here we must account for the different specific heats of water and ceramic. We have $q_w + q_{cup} = 0$ with a common final temperature T; thus

$$m_w c_w (T - T_w) + m_{cup} c_{cup}(T - T_{cup}) = 0$$

$$T = \frac{m_w c_w T_w + m_{cup} c_{cup} T_{cup}}{m_w c_w + m_{cup} c_{cup}}$$

$$= \frac{(150. \text{ cal}/^\circ\text{C})(100. \ ^\circ\text{C}) + (38 \text{ cal}/^\circ\text{C})(22 \ ^\circ\text{C})}{(150. + 38) \text{ cal}/^\circ\text{C}} = 84.2 \ ^\circ\text{C}$$

Note that the result for T is now linear in the "heat capacity fraction." Because the specific heat of ceramic is so much lower than that of water, the final temperature is much closer to the initial hot water temperature.

13. Although here we are after an atomic mass M, the First Law governs the outcome through $q_m + q_w = 0$. For the heat capacity of the metal, we use $C = mc = (m/M)(Mc) = (m/M)\overline{C}$, where the Law of Dulong and Petit states that $\overline{C} = 6$ cal/K mol, and K and $^\circ$C are equivalent. We have

$$(m_m/M)\overline{C}_m(T - T_m) + m_w c_w (T - T_w) = 0$$

$$M = -\frac{m_m \overline{C}_m(T - T_m)}{m_w c_w (T - T_w)}$$

$$= -\frac{(20.0 \text{ g})(6 \text{ cal}/\text{K mol})(26.4 - 200) ^\circ\text{C}}{(75. \text{ cal}/^\circ\text{C})(26.4 - 22.0) \ ^\circ\text{C}}$$

$$= 63 \text{ g}/\text{mol}$$

The metal is most likely copper, whose experimental $\overline{C}$ is 5.84 cal/K mol.

15. If the pump temperature is rising, then the First Law states that $\Delta E = q + w > 0$, or that w is being converted to E, since no external heat source is applied. The fact that the pump gets warm implies that $q < 0$; combined with $\Delta E > 0$, this means that the pump is unable to conduct the heat away as quickly as the work is being added. Presumably, if the pump strokes were more and more gradual, the warming per stroke would be less and less, and in the limit ΔE would be zero, and the process would be isothermal. The exercise considers the opposite, adiabatic limit, for which $q = 0$ and $\Delta E = w$. Here the solution is easier because we know the final volume, allowing the work to be readily computed from Equation 10.7,

$$w = -P_2(V_2 - V_1) = -(6.00 \text{ atm})[(1.2 - 0.2) \text{ L}](24.22 \text{ cal}/\text{L atm}) = 145 \text{ cal}$$

and, since $\Delta E = C_V \Delta T$, we can easily find ΔT. From Table 10.4, $\overline{C}_V = \overline{C}_P - R = 6.93 - 1.99 = 4.94$ cal/K mol, and from the ideal gas law for state 1, $n = 0.0508$ mol, making $C_V = (0.0508 \text{ mol})(4.94 \text{ cal}/\text{K mol}) = 0.251$ cal/K. Thus

$$\Delta T = \Delta E/C_V = w/C_V = 145.3 \text{ cal}/(0.251 \text{ cal}/\text{K}) = 579 \text{ K}$$

or a final temperature of $594°C$. In the actual hand pump, the air is not trapped, and there is no insulation, giving a much more modest ΔT. In a diesel engine, however, the high final T ignites the fuel.

17. Equation 10.22 defines enthalpy as $H = E + PV$. Then

$$\begin{aligned}
\Delta H &= H_2 - H_1 \\
&= E_2 + P_2 V_2 - (E_1 + P_1 V_1) \\
&= E_2 - E_1 + (P_2 V_2 - P_1 V_1) \\
&= \Delta E + \Delta(PV)
\end{aligned}$$

For one mole of an ideal gas or a mixture of ideal gases, $\Delta(PV) = \Delta(RT) = R\Delta T$, leading to $\Delta H = \Delta E + R\Delta T$, and $\Delta H > \Delta E$ for a heating process. These two energies can be equal only for an isothermal process in an ideal gas. When the heating is isobaric, $\Delta(PV) = P\Delta V = -w = R\Delta T$, and ΔH is greater than ΔE by $-w$, the amount of work done by the gas. When the heating is isochoric, on the other hand, $\Delta(PV) = V\Delta P = R\Delta T$, and there is no P-V work done, but ΔH is still greater than ΔE by the amount of work that can potentially be done if the system with its increased pressure is vented to the surroundings isothermally. This potential energy we will later call the *free energy* (see Chapters 11 and 12).

19. Since the path from ground to stratosphere is not specified, we are saved from having to calculate q and w—there is no way to obtain them without that knowledge. But we can calculate ΔE and ΔH, the initial and final states being specified by having solved the example. If the ideal gas law is followed, furthermore, all that matter are T and n; from the example, $T_1 = 7°C$, and $T_2 = -43°C$, and using the ideal gas law in state 1, $n = 41.8$ mol. Regarding He as an ideal monatomic gas, Equations 10.19 and 10.24 give us

$$\Delta E = \tfrac{3}{2} n R \Delta T = \tfrac{3}{2}(41.8 \text{ mol})(1.9872 \text{ cal}/\text{K mol})(-50\,°C) = -6230 \text{ cal}$$

$$\Delta H = \tfrac{5}{2} n R \Delta T = \tfrac{5}{3}(-6230 \text{ cal}) = -10380 \text{ cal}$$

You may intuitively realize that these First-Law quantities do not provide a full description of the system; they miss the fabulous gas expansion at stratospheric pressures. This will be described by the so-called *free energy* in Chapter 12.

21. Following Example 10.4, the calibration data give us the heat capacity C of the calorimeter,

$$C = \frac{q_{elec}}{\Delta T} = \frac{V^2 t}{R\Delta T} = \frac{(115 \text{ J}/\text{C})^2 (120.\text{ s})}{(100.\text{ J s}/\text{C}^2)(3.833\,°C)}(1 \text{ cal}/4.184 \text{ J}) = 989.6 \text{ cal}/°C$$

Here V is voltage and R resistance, with units reduced to J and C. Then $\Delta E° = q_V$ is

$$\Delta E° = -\frac{(989.6 \text{ cal}/°C)(9.617\,°C)}{1.505 \text{ g}}(44.053 \text{ g}/\text{mol})(1 \text{ kcal}/10^3 \text{ cal}) = -278.6 \text{ kcal}/\text{mol}$$

To convert ΔE to ΔH using Equation 10.28, we need the balanced combustion equation to find Δn_{gas}, $CH_3CHO(l) + \tfrac{5}{2}O_2(g) \rightarrow 2CO_2(g) + 2H_2O(l)$. Thus $\Delta n_{gas} = 2 - 2.5 =$

−0.5 mol, and

$$\Delta H^{\circ} = \Delta E^{\circ} + \Delta n_{gas}RT$$
$$= -278.6 \text{ kcal} + (-0.5 \text{ mol})(1.9872 \text{ cal}/\text{K mol})(298.15 \text{ K})(1 \text{ kcal}/10^3 \text{ cal})$$
$$= -278.9 \text{ kcal}$$

The specific reaction heat is $(-278,900 \text{ cal/mol})/(44.053 \text{ g/mol}) = 6331 \text{ cal/g}$. Note that for the bomb calorimeter there is no work, $w = 0$, but ΔH measures the heat evolved at constant pressure (e.g., burning CH_3CHO in air). The correction term is actually $-w$.

23. The balanced chemical equations and their heats are:

ΔH° (kcal/mol)

1.	$CH_4(g) + 2O_2(g) \rightarrow CO_2(g) + 2H_2O(g)$	−191.6
2.(× −2)	$H_2O(l) \rightarrow H_2O(g)$	10.52
3.	$CH_4(g) + 2O_2(g) \rightarrow CO_2(g) + 2H_2O(l)$	−212.65

We have applied Hess's Law, which allows us to manipulate the ΔHs in the same way we manipulated the chemical equations, i.e.,

$$\Delta H_3 = \Delta H_1 - 2\Delta H_2$$

giving the result above. An enthalpy-level diagram is given on the next page; the setting of an enthalpy zero is arbitrary. The percent increase in efficiency is therefore $[(212.65 - 191.6)/191.6] \times 100\% = 11\%$. This may not seem like a big advantage, but it can save a typical household in northern climes hundreds of US\$ per year as this is written; and as natural gas prices increase, the advantage grows.

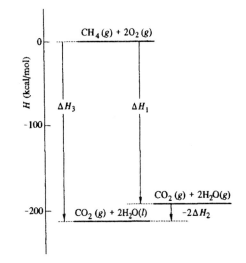

25. Given that reaction 1 is the most exothermic, we look for a way to generate it from reactions 2 and 3. We see that reactions 2 and 3 each contain a mole of reagent needed for reaction 1, and when we add these, both the NH_3 and the NH_4Cl cancel out, and reaction 1 is recovered. This means that, if reactions 2 and then 3 (or 3 and then 2) are run in succession, the result will be the products of reaction 1 *plus* an extra mole of NH_3 (or NH_4Cl), and the heat will be the same as reaction 1, that is

$$\Delta H_1 = \Delta H_2 + \Delta H_3$$

The corresponding enthalpy diagram is shown on the right; we show the actual reaction heats, which are calculated in Exercise 32. The enthalpy zero is arbitrary. In order to represent the cascading reactions, we need to add a mole of either NH_3 or NH_4Cl to both

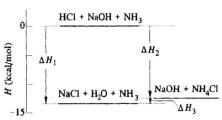

sides of reaction 1.

27. This is a redox addition reaction (see Chapter 5), for which, if the elements begin in their standard states, the heat of reaction is just the stoichiometric coefficient of the adduct multiplied by its heat of formation. Using Table 10.1,

$$\Delta H^{\circ} = 2\Delta H_f^{\circ}(NaCl(s)) = 2(-98.268)$$
$$= -196.536 \text{ kcal/mol}$$

Using heats of formation sets the zero of enthalpy at the elements in their standard states; an enthalpy diagram is shown on the right.

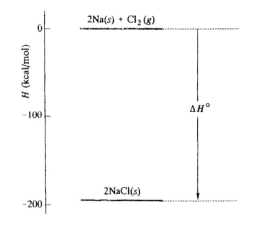

29. Because the product of $2O_3(g) \rightarrow 3O_2(g)$ is an element in its standard state, Equation 10.30 and Table 10.1 give us $\Delta H^{\circ} = -2\Delta H_f^{\circ}(O_3(g)) = (-2)(34.11) = -68.22$ kcal/mol, or -34.11 kcal/mol O_3. This exothermic reaction heats the stratosphere. See Chapter 15 for a discussion of the ongoing catalytic destruction of the ozone layer.

31. Table 10.1 and Equation 10.30 give
$$2NaHCO_3(s) \rightarrow Na_2CO_3(s) + CO_2(g) + H_2O(g)$$
$$\Delta H^{\circ} = (-2)(-227.25) + (1)(-270.24) + (1)(-94.052) + (1)(-57.796)$$
$$= +32.41 \text{ kcal/mol}$$

33. The sulfate ion is a spectator; cancelling it gives us the net ionic equation and, using data from Table 10.2, its heat
$$Zn(s) + Cu^{2+}(aq) \rightarrow Zn^{2+}(aq) + Cu(s)$$
$$\Delta H^{\circ} = (-1)(0) + (-1)(15.48) + (1)(-36.78) + (1)(0) = -52.26 \text{ kcal/mol}$$

35. The net ionic equation enables us to use Tables 10.1 and 10.2 to find ΔH° as follows:
$$MnO_4^- + 5Fe^{2+} + 8H^+ \rightarrow Mn^{2+} + 5Fe^{3+} + 4H_2O$$
$$\Delta H^{\circ} = (-1)(-129.4) + (-5)(-21.3) + (-8)(0) + (1)(-52.76) + (5)(-11.6) + (4)(-68.315)$$
$$= -148.1 \text{ kcal/mol}$$

37. For this simple bond-switching reaction, Equation 10.32 and Table 10.3 give us
$$Na(g) + Cl\text{—}Cl(g) \rightarrow Na^+Cl^-(g) + Cl(g)$$
$$\Delta H^{\circ} = D_0(Cl\text{—}Cl) - D_0(Na^+Cl^-)$$
$$= 58 - 98 = -40. \text{ kcal.}$$

The gas-phase atoms $Na(g)$ and $Cl(g)$ are defined to lie at the zero of enthalpy, and are therefore not included in the calculation. Bond lengths are estimated from Figures 4.12 and 4.13 as
$$r_e(Cl_2) = 2(0.99 \text{ Å}) = 1.98 \text{ Å}$$
$$r_e(NaCl) = (0.85 + 1.50)\text{Å} = 2.35 \text{ Å}.$$
As shown in the graph at the right, ΔH° is measured from the zero-point vibrational

level of Cl_2 to that of NaCl; thermal excitation
at 298K provides a negligibly small correction.
Both of these adjustments are included when
tables of bond energies are used. The general
properties of covalent and ionic bond potential
energy curves have been discussed in the
solution to Exercise 6.3. In this reaction, the
nascent NaCl molecule has been shown to be
highly vibrationally excited in molecular beam
experiments; that is, much of ΔH° is initially
deposited in vibration.

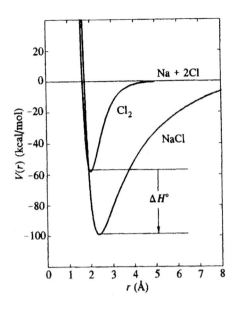

39. Following the method of Example 10.7, we write the formation reaction and find its heat
using Equation 10.32 and Table 10.3:
$$\tfrac{1}{2}N_2(g) \;+\; H_2(g) \;\rightarrow\; NH_2(g)$$
$$\Delta H^\circ = (1)\Delta H_a(N) + (2)\Delta H_a(H) + (-2)D_0(N\!-\!H)$$
$$= 113.0 + 2(52.1) - 2(93) = 31 \text{ kcal/mol}$$

This is considerably lower than the literature
44 kcal, owing to the greater strength of the
$H_2N\!-\!H$ bond (107 kcal) relative to $HN\!-\!H$
(94) or $N\!-\!H$ (78). The enthalpy diagram is at
the right.

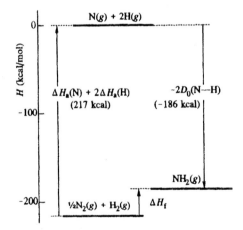

41. Setting up both caculations, using Table 10.1 and Equation 10.30 in the first and Table
10.3 and Equation 10.32 in the second, we have
$$4NH_3(g) \;+\; 5O_2(g) \;\rightarrow\; 4NO(g) \;+\; 6H_2O(g)$$
$$\Delta H^\circ = (-4)(-11.02) + (-5)(0) + (4)(21.57) + (6)(-57.796)$$
$$= -216.43 \text{ kcal/mol from } \Delta H_f^\circ \text{s}$$
$$\Delta H^\circ = (12)(93) + (5)(118) + (-4)(150) + (-12)(110)$$
$$= -214 \text{ kcal/mol from } D_0 \text{s}$$

The first is more accurate, due in this case to the accumulation of rounding errors in the
bond energies. As the last two exercises showed, if there were hydride radicals
involved, the results would not agree so well. You are encouraged to draw an enthalpy
level diagram comparing the energetics of these two approaches; this would be analogous
to Figure 10.7.

43. CO_2 being linear, we use the classical Equation 10.33a for both it and HCl, giving

$$\bar{C}_P(HCl) = \left(3 \cdot 2 - \tfrac{3}{2}\right)R = 4.5R = 4.5(1.9872 \text{ cal}/K \text{ mol}) = 8.942 \text{ cal}/K \text{ mol}$$

$$\bar{C}_P(CO_2) = \left(3 \cdot 3 - \tfrac{3}{2}\right)R = 7.5R = 14.904 \text{ cal}/K \text{ mol}$$

Both of these are larger than the values in Table 10.4, 6.96 and 8.87, by about R and $3R$, respectively. As discussed in the text, the overestimates stem from underpopulated excited vibrational DFs for which $h\nu > k_B T$. Since classically each vibrational DF yields a contribution of R to $\bar{C}_P$, for HCl the vibrational contribution is essentially zero, while for CO_2 it is only R instead of $4R$. At 298K, $k_B T = 207 \text{ cm}^{-1}$; comparing this to the vibrational wavenumbers of Table 8.3 [HCl, $\bar{\nu} = 2990 \text{ cm}^{-1}$; CO_2, $\bar{\nu}_a = 2349$, $\bar{\nu}_s = 1388$, and $\bar{\nu}_b = 667 \text{ cm}^{-1}$ (twice)], we see that only for the bending vibrations of CO_2 is $h\nu$ comparable to $k_B T$. This last is the chief origin of the vibrational contribution to $\bar{C}_P(CO_2)$. The vibrational heat capacity for a given DF may be computed exactly in the harmonic oscillator model using a formula first derived by Einstein in 1907, $\bar{C} = R[(x/2)/\sinh(x/2)]^2$, where $x = hc\bar{\nu}/k_B T$. For CO_2 this yields a vibrational contribution of 1.90 cal/K mol coming mainly from $\bar{\nu}_b$, and a total heat capacity of $1.90 + 3.5R = 8.85$ cal/K mol, in good agreement with the tabulated value.

45. The Law of Dulong and Petit, $\bar{C} \approx 6$ cal/K mol for metals, is discussed in Exercise 13 and Section 10.7. The specific heat is $c = \bar{C}/M$; for Cr(s) this is (6 cal/K mol) / (52.0 g/mol) = 0.115 cal/K g, higher than the literature value by 7.5%. Since c varies inversely as the atomic mass M, $c(Al) > c(Cr)$, and Al will absorb more heat per unit mass than Cr for a given ΔT.

47. Using $q = C\Delta T = n\bar{C}\Delta T$ (Equation 10.10) with $n = (0.350 \text{ L})/(22.414 \text{ L/mol}) = 0.01562$ mol, and $\bar{C}_P - \bar{C}_V = R$ (Equation 10.26), we can use $\bar{C}_P(CH_4(g)) = 8.439$ cal/K mol from Table 10.4 to find

$$q_P = n\bar{C}_P\Delta T = (0.01562 \text{ mol})(8.439 \text{ cal}/K \text{ mol})(600 - 0)^\circ C = 79.1 \text{ cal}$$

$$q_V = n\bar{C}_V\Delta T = n(\bar{C}_P - R)\Delta T = 60.5 \text{ cal}$$

These results neglect the temperature dependence of $\bar{C}_P$, thus underestimating the heat needed, since $\bar{C}_P$ always increases with T as the vibrations begin to absorb heat.

49. This exercise is similar to Exercise 47. Here $n = (1000 \text{ g})/(44.0 \text{ g/mol}) = 22.7$ mol, and from Table 10.4 $\bar{C}_P(CO_2(g)) = 8.87$ cal/K mol. We have

$$q_P = n\bar{C}_P\Delta T = (22.7 \text{ mol})(8.87 \text{ cal}/K \text{ mol})(-78 - 25)^\circ C = -20.8 \text{ kcal}$$

$$q_V = n\bar{C}_V\Delta T = n(\bar{C}_P - R)\Delta T = -16.1 \text{ kcal}$$

Less refrigeration is required at constant volume. As in Exercise 47, the dependence of $\bar{C}_P$ on T has been neglected; in the present system, this is a better approximation, because only the already small contribution from vibration (see Exercise 43) is lost on cooling.

51. The numerical results for this Exercise are in the Answer Key; see also Figure 10.8.

The calculations center around the variable $x = hc\bar{\nu}/k_BT = (1.4388\ \text{cm K})\bar{\nu}/T$. From Table 8.3, $\bar{\nu} = 565\ \text{cm}^{-1}$ for Cl_2, giving $x = 2.73$ at 298K. The partition function q for vibration (not to be confused with heat q) is just the sum of the Boltzmann factors for each state

$$q = e^0 + e^{-x} + e^{-2x} + e^{-3x} + \ldots = 1 + 0.0654 + 0.0043 + 0.0003 + \ldots$$

Here using only the first three terms yields $q = 1.070$, correct to 4S. The exact result (see Exercise D.6) is $q = (1 - e^{-x})^{-1} = 1.0699\ldots$ Then the fractional populations $f(v)$ $= e^{-vx}/q$ are readily computed, along with $E_{\text{vib}} = N\sum_v hc\bar{\nu}vf(v) = Nhc\bar{\nu}\sum_v vf(v)$, by evaluating each term and summing. The energy factor in kcal/mol is computed as

$$N_A hc\bar{\nu} = (6.022 \times 10^{23}\ /\text{mol})(6.626 \times 10^{-34}\ \text{J s})(2.998 \times 10^8\ \text{m/s})(56500\ \text{m}^{-1})$$

$$\times \left(\frac{1\ \text{cal}}{4.184\ \text{J}}\right)\left(\frac{1\ \text{kcal}}{1000\ \text{cal}}\right)$$

$$= 1.615\ \text{kcal/mol}$$

Then

$$E_{\text{vib}} = 1.615[0 \cdot 0.9349 + 1 \cdot 0.0611 + 2 \cdot 0.0040 + \ldots] \approx 0.112\ \text{kcal/mol}$$

Note that the ground state does not contribute to E_{vib}. The exact result (see Exercise A.1) is $E_{\text{vib}} = Nhc\bar{\nu}(e^x - 1)^{-1} = 0.1129\ldots$ kcal/mol. At 298K the classical vibrational energy is $RT = 0.592$ kcal/mol; thus the vibrational DF is not able to store the full thermal energy, reducing the vibrational heat capacity. See Exercise 43 for an exact formula for the vibrational heat capacity. The exact harmonic oscillator results can be obtained to any desired accuracy by adding more terms to the sums; to get results pertinent to the "real" Cl_2 molecule requires accounting for the actual, declining spacing between vibrational states as v increases (see Figure 8.6).

53. Writing the balanced formation reaction and ΔC_P beneath,

$$C(s) \quad + \quad 2H_2(g) \quad \rightarrow \quad CH_4(g)$$

$$\Delta C_P = (-1)(2.038) + (-2)(6.889) + (1)(8.439) = -7.377\ \text{cal/K mol}$$

Then

$$\Delta H^\circ_{f,T} = -17.88 - 0.007377(T - 298)$$

At $T = 230$K this yields $\Delta H^\circ_{f,230} = -17.38$ kcal/mol. H_R falls more rapidly with T than does H_P, making the reaction less exothermic. You are encouraged to construct an enthalpy level diagram analogous to Figure 10.9 that illustrates this.

11

Spontaneity of Chemical Reactions

Your Chapter 11 GOALS:

- Understand the "need" for entropy to explain endothermic, spontaneous reactions
- Compare and contrast reversible and irreversible processes
- Use the expansion of a gas to determine the relationship between reversible and irreversible work
- Explain the significance of the Carnot cycle
- State the mathematical definition of the change in entropy for a process
- Define what it means for an integral to be path-dependent
- State the Second Law of thermodynamics
- Use configuration and arrangement analysis to define entropy and spontaneous processes
- Use absolute entropies to calculate the entropy change of a reaction
- Define the Gibbs free energy
- Recognize the competition between enthalpy and entropy in spontaneity considerations
- Determine the effect of temperature on spontaneity
- Understand the relationship between ΔG and work

Chapter 11 KEY EQUATIONS:

- 11.1, 11.2, 11.6, 11.8, 11.10, 11.11, 11.12, 11.13, 11.14, 11.17, 11.19, 11.24, 11.25

Overview

Our Chapter 10 introduction mentioned the fact that heat (or thermal energy) flows *spontaneously* from hot bodies into cold bodies, allowing us to say $q_{hot} = -q_{cold}$. If the reverse were true, equilibrium would never be reached, and, as an example, an ice cube that is placed on a counter top would get colder and colder while the counter top would get hotter and hotter and, eventually, ignite! The quantity called entropy quantifies spontaneity (melting ice) and impossibility (igniting counter tops!). The difficult conceptual part is that, like "energy", there is no precise *thermodynamic* definition of entropy itself, and entropy is defined in terms of its *change* in Equation 11.8. See also Equation 11.13. The change in entropy, like the change in internal energy, is monitored by *measurable* quantities such as q and T.

While enthalpy and the topic of thermochemistry are encountered early in typical introductory chemistry textbooks, entropy is saved for much later chapters. For example, a student could take a full year of high school chemistry and *never* even hear the word entropy. Even in freshman general (college) chemistry classes, entropy is presented in conjunction with ΔG (Equation 11.24) and rarely given specialized treatment. If this has been the case with you, which is likely, note that your previous teachers were not necessarily being remiss in their duties. As you have become keenly aware, entropy is abstract. Nonetheless, it is every bit as fundamental as enthalpy; enthalpy has one law of thermodynamics that specifically addresses it, whereas entropy has two! For now it is acceptable to answer the question 'What is entropy?' with one of two answers: i) Entropy measures the ability of a system to undergo spontaneous change, or ii) Entropy is a measure of the disorder of a system that is based on the probability of the system being in a particular arrangement. The first answer is the thermodynamic answer (Equation 11.8), and the second answer is the statistical-mechanics answer (Equation 11.14).

Exercises 1, 2, and 3 compare the differences in the work of reversible and irreversible processes via Equations 11.1 and 11.2. Exercise 2 is an attempt to illustrate the importance of the stair-step curve in the plot of Figure 11.3. The work associated with the expansions is -135.6 cal and -180.8 cal, respectively. The sum of these ought to be compared to the values of $w_{rev} = -376$ cal and $w_{irrev} = -271$ cal from Section 11.2. For Exercise 4 Equation 9.13 gives V_1 and P_2, and V_2 can be obtained from the last line of Equation 11.5. You'll need to be on your toes with the subscripts because V_3 represents a *final* volume in Equation 11.5. Exercise 5(c) is another one of those answers that merits repeated reading to ensure comprehension. The concepts addressed are crucial. Now, while Newton and Boltzmann were vanguards of science, the lesser known Carnot was as well, and his findings were of great significance – once they were understood! Exercises 6 through 9 contain two algebraic dilemmas and a rather simple Exercise. While these Exercises can help you to learn about the Carnot cycle, this is, no doubt, new to you, and a significant amount of thought must be given to what is being considered by Carnot.

Typically, math is a stumbling block, but entropy is so odd that many students find that Exercises which involve its calculation are easier than any qualitative question such as 'What is entropy?'. See above. Learning is a long process, and there is no shame in gathering as many pebbles of knowledge (that is, integrating your knowledge!) as you can – whenever you can. If manipulating entropy equations is your first step to understanding this concept, then so be it. Time is on your side. Exercise 10 begins the quantitative look at entropy, and it contains the very important lesson that ΔS is a state function, meaning that *its value will be the same for either a reversible or an irreversible path*. This does *not* mean, however, that $q_{rev}/T = q/T$. Use the fact that $q_{rev} = -w_{rev}$ and $q_{irrev} = -w_{irrev}$ along with Equations 11.1 and 11.2 to show this. For Exercises 12 and 13 be sure to use the appropriate heat capacity, using Equation 10.26 as necessary. For Exercise 14 note that the condensation is reversible and isothermal, allowing the use of Equation 11.11 once q_P has been determined. Even if you are not assigned Exercise 16, it will be helpful to read the Exercise in order to assist you with your prediction for Exercise 23(e). Exercise 17 is another one of those computationally simple Exercises with profound implications. (We have discussed Exercise 15 of Chapter 1 and Exercise 21 of Chapter 8 in such a manner.) Exercise 17 here is related to Exercise 28 of this Chapter and, therefore, to the operation of a pH electrode, an instrument of great significance all over the world.

Above, we said that there is no precise *thermodynamic* definition of entropy. There is, however, a precise *statistical thermodynamic* definition of entropy, Equation 11.14. Both thermodynamics and statistical thermodynamics describe macroscopic properties of systems. Unlike thermodynamics, however, statistical thermodynamics acknowledges the existence of atoms and derives macroscopic properties from the probabilistic behavior they exhibit. For this reason the monumental Equation 11.14 is a link between the classical and atomistic, probabilistic world. Appendix D gives a short introduction to this complex area of study, and if you examine Exercise 5 of this Appendix, you can see that the (classical) thermodynamic formula Equation 11.11 is equivalent to the much more fearsome looking Equation D.32. This strategy is akin to what we saw with the Bohr Correspondence Principle in Exercises 26 and 27 of Chapter 3, where classical equations must be obtained as limiting cases of the quantum equations. Boltzmann's work showed that entropy is only about the probabilities of a large collection of molecules and that a single atom does not have "entropy". It is his approach that yields the common, nearly pop-culture, correlation of "disorder" and entropy. Exercises 19, 20, and 21 all require the use of Equation D.3, and all carry the take-home message that the most likely configuration is the one with the most arrangements, or disorder. For Exercise 20 the "interesting" configuration has a total of 60 possible arrangements and $W_2 = 3^6 = 729$. Equation 11.15 is then needed with the ln ratio of 729/1.

Qualitative questions about the entropy of substances (Exercise 22) or those seeking predictions of $\Delta S°$ (Exercises 23) are simple to answer, relying on relationships such as $S°$(solid)

$< S°$(liquid) $< S°$(gas) and $\Delta n_{gas} <$ or > 0. In particular, however, note that the simplicity of Exercise 22 belies its significance. Whereas absolute values for internal energy or enthalpy depend on choosing an *arbitrary* zero, the Third Law of thermodynamics yields absolute entropies of substances due to a zero that is *not* arbitrary. Above we have indicated that there exists a thermodynamic and a statistical thermodynamic interpretation of entropy. The thermodynamic concept of entropy posits that at $T = 0$ K $\Delta S = 0$ for any possible process, and the statistical thermodynamic concept of entropy posits that $W = 1$ at the absolute zero of temperature. (These ideas are presented just before Equation 11.17.) These two concepts of entropy both come to the same conclusion: $S_0 = 0$ for a perfect crystal.

Equation 11.19 is needed to solve Exercises 24, 29, 30, 32, and 33. As discussed briefly above, these allow you to work with entropy values and to incorporate them into Equation 11.24 even if you are not yet sure what entropy really is. Thus, we arrive at the final topic of this Chapter – the Gibbs free energy ΔG. Gibbs is a scientific giant in league with Newton and Einstein, and Equation 11.24, now entrenched in chemical thought, is a testament to his genius. Neither enthalpy alone nor entropy alone controls a chemical reaction, and it was Gibbs that recognized and quantified the competition between them in the form of ΔG. Further, the idea of ΔG as a driving force for a reaction relies on equating the free energy with the maximum work w_{rev} that a system can perform (Exercise 26). ΔG, which may be new to a number of you, will appear frequently throughout the remainder of *University Chemistry*. ΔG explains the spontaneity of dilutions (Exercise 28), influences to extent to which reactions proceed via its relationship with the equilibrium constant K (see Chapter 12), dictates phase change behavior (Chapter 14), and addresses the reason why some reactions experience a change in the spontaneity as the temperature changes (see Equation 11.25 and Figure 11.8). Strictly speaking, since it is derived from Equation 11.20, ΔG is *not* a fundamental idea within the framework of the three traditional Laws of thermodynamics. Still, ΔG is a very convenient summation of the Laws of thermodynamics, and chemists use it frequently and effectively. In this sense it is fundamental because it measures the capacity for a system to undergo spontaneous change.

Exercises 26 through 29 deal with isothermal conditions. Thus, since $\Delta H = C_p \Delta T$, $\Delta H = 0$ for these problems, and $\Delta G = - T\Delta S$ in each case. Exercise 30, which is the continuation of Exercise 31 of Chapter 10, along with Exercises 31, 32, and 33 are simple applications of Equation 11.25 via Equations 10.30 and 11.19. Note that Exercises 30, 31, and 32 yield positive, and real, T_{ta}'s. Exercise 33 yields a negative T_{ta}, which is impossible because the unit is K, not °C.

1. According to Equations 11.1 and 11.2, isothermal work for an ideal gas depends only on n, T, and volume or pressure ratios, in the case that $P_{ext} = P_2$. Thus the actual pressures or volumes need not be known, although for simplicity you could easily assume, say, $P_2 = 1$ atm, and proceed to solve the problem. For our case, $V_2 = 3V_1$, the reversible work is

$$w_{rev} = -nRT\ln\left(\frac{V_2}{V_1}\right)$$
$$= -(2\,\text{mol})(1.9872\,\text{cal}/\text{K mol})(298\,\text{K})\ln 3 = -1301\,\text{cal}\ (-5440\text{J})$$

The work ratio *for the same final state* is

$$\frac{w_{rev}}{w_{irrev}} = \frac{-nRT\ln 3}{-nRT\left(1 - \frac{1}{3}\right)} = \tfrac{3}{2}\ln 3 = 1.65$$

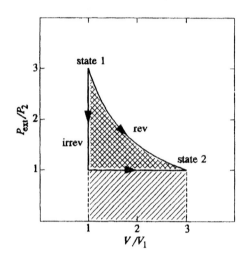

Comparing this to the doubled volume discussed in the text, the reversible advantage is greater; for a volume ratio r, the work ratio is $[r/(r-1)]\ln r$, which grows monotonically as r increases. The advantage is evident in comparing the plot on the right to Figure 11.3.

3. For the one-stage compression $P_{ext} = P_1 = 2$ atm, while for the two-stage we assume $P_2 = 1.50$ atm. We use Equation 11.2 combined with Boyle's Law to find

$$\text{For 1 stage:}\quad w = nRT\left(\frac{P_1}{P_3} - 1\right)$$
$$= (542.5\,\text{cal})(2.00 - 1) = 543\,\text{cal}$$
$$\text{For 2 stages:}\quad w = nRT\left(\frac{P_2}{P_3} + \frac{P_1}{P_2} - 2\right)$$
$$= (542.5\,\text{cal})(1.50 + 1.333 - 2) = 452\,\text{cal}$$

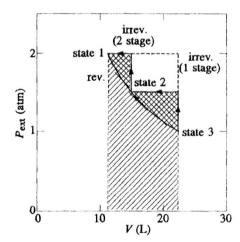

The work of reversible compression is the negative of expansion, $+376$ cal, and we see that the two-stage compression is closer to the reversible limit. The $P\text{-}V$ diagram is given on the right. Combined with Exercise 2, the calculations are an indication that work output is maximized in expansion and input minimized in compression when carried out reversibly.

5. This is an irreversible adiabatic expansion. Because V_2 is given, this Exercise is considerably easier than Exercise 10.14.b, our previous exposure to irreversible adiabatics. We first calculate w, then, using $\Delta E = w = C_V \Delta T$, T_2.

a. $w = -P_{ext}\Delta V = -(1.00 \text{ atm})(36.9 - 24.6)\text{L}(24.22 \text{ cal/L atm})$
$$= -298 \text{ cal} \ (-1250 \text{ J})$$

b. From Table 10.4, $\bar{C}_P = 6.961$ cal/K mol for $N_2(g)$. Then

$C_V(T_2 - T_1) = w$
$$T_2 = T_1 + \frac{w}{C_V} = 300 \text{ K} + \frac{-298 \text{ cal}}{(2 \text{ mol})(6.961 - 1.987) \text{ cal}/\text{K mol}}$$
$$= 270 \text{ K}$$

and $P_2 = nRT_2/V_2 = 1.20$ atm.

c. Because a reversible expansion produces more work, there is more cooling. Therefore $T_2 < 270$K, and consequently $P_2 < 1.20$ atm for the given V_2. This work can be computed using the $PV^\gamma = $ constant relation found in Exercise 4. For N_2, the classical formula Equation 10.33 without vibration predicts $\gamma = \frac{7}{5}$, very close to what is obtained using Table 10.4. We have

$$-w = \int_{V_1}^{V_2} P\,dV = P_1 V_1^\gamma \int_{V_1}^{V_2} \frac{dV}{V^\gamma} = \frac{P_1 V_1^\gamma}{1-\gamma}\left(V_2^{1-\gamma} - V_1^{1-\gamma}\right)$$

$$= \frac{nRT_1}{\gamma - 1}\left(1 - \left(\frac{V_1}{V_2}\right)^{\gamma-1}\right)$$

$$= 2.5(1192 \text{ cal})\left(1 - \left(\frac{24.6}{36.9}\right)^{0.4}\right) = 446 \text{ cal}$$

This is, as expected, larger than $-w$ of part (a). Proceeding as in part (b), we find $T_2 = 255$ K and $P_2 = 1.13$ atm, less than those of part (b) also as expected.

7. All we need here is the starkly simple result of Carnot, Equation 11.6. In kelvins our temperatures are 288K and 418K, yielding $\varepsilon = 1 - (288/418) = 0.311$. To increase ε we can either raise T_h, lower T_c, or both. T_c is an environmental variable, not readily controlled, so running the engine hotter is the only alternative. Higher operating T, however, causes new problems, mainly thermal expansion leading to cracks in the engine block, decomposition of lubricants, and failure of elastomer seals on shafts and valve seats. 418K is a compromise between efficiency and long engine life.

9. The data and paths given allow us to characterize all four states; note that the isothermal stages remain, but the adiabatic stages have been eliminated in favor of isochoric cooling / heating. For simplicity we assume an atomic ideal gas. We have

State	T(K)	P(atm)	V(L)	Path	w(cal)	q(cal)	
1	600	10.0	4.92				
				isoth exp	−596	+596	(Equation 11.2)
2	600	5.0	9.85				
				isoch clg	0	−894	(Equation 10.27)
3	300	2.5	9.85				
				isoth cmp	+596	−596	
4	300	5.0	4.92				
				isoch htg	0	+894	

From Equation 11.13, the net work, and thus the efficiency, is zero, implying that the area enclosed by the cyclic path is also zero, as the P-V diagram on the right illustrates. This cycle requires, however, *instantaneous* pressure release on expansion and application on compression. Any finite duration, as suggested by the dashed lines, implies a slight amount of reversibility creeping into the cycle, and thus this limit is as unreal as Carnot's.

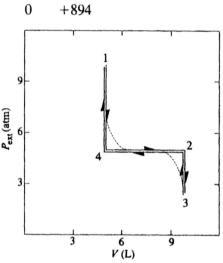

11. Boyle's Law $P_1V_1 = P_2V_2$ is equivalent to $P_1/P_2 = V_2/V_1$. Replacing the volume ratio in Equation 11.11 with the pressure ratio gives $\Delta S = nR\ln(P_1/P_2)$. Thus, isothermal compression of a gas yields a negative entropy change:

$$\Delta S = (0.250 \text{ mol})(1.9872 \text{ cal/K mol}) \ln(1.00/6.00) = -0.890 \text{ cal/K } (-3.72 \text{ J/K})$$

13. At constant pressure Equation 11.12, with C_P replacing C_V, gives

$$\Delta S = C_P\ln\frac{T_2}{T_1} = (5.00 \text{ mol})(8.6 \text{ cal/K mol}) \ln\frac{373\text{K}}{623\text{K}} = -22.1 \text{ cal/K } (-92 \text{ J/K})$$

The heat capacity of steam increases slowly between 100°C and 350°C, reaching 8.75 cal/K mol; this yields a slightly greater drop in S.

15. We use the isothermal reversible expansion Equation 11.11 to find ΔS, with $n = 2.009$ mol from the ideal gas law with $V = V_1 = 2.196$ L (134 in³). Assuming two identical gas containers, we have

$$\Delta S = nR\ln\frac{V_2}{V_1} = (2.009 \text{ mol})(1.9872 \text{ cal/K mol})\ln 2 = 2.77 \text{ cal/K } (11.58 \text{ J/K})$$

17. Conservation of moles HCl implies $c_1V_1 = c_2V_2$ and
$$V_2 = (10.0 \text{ mL})(1.00 \text{ M})/0.0500 \text{ M} = 200. \text{ mL}$$
Assuming additivity, 190. mL of water must be added. Equation 11.11 then gives
Lastly, $c_1V_1 = c_2V_2$ rearranges to $V_2/V_1 = c_1/c_2$. Direct substitution into Equation 11.11

$$\Delta S = nR\ln\frac{V_2}{V_1} = (0.0100\,\text{mol})(1.9872\,\text{cal/K mol})\ln\frac{200.\,\text{mL}}{10.0\,\text{mL}} = 0.0595\,\text{cal/K}\ \ (0.249\,\text{J/K})$$

gives $\Delta S = nR\ln(c_1/c_2)$.

19. From the formulas given in
Section 11.4, particularly in
Figure 11.7, we find that
$W_1 = 1^N = 1$, and $W_2 = 2^N$.
For $N = 4$, $W_2 = 16$. The
number of arrangements for a
given configuration is $W_{LR} =$
$N!/(L!R!)$; for example

$$W_{13} = \frac{4!}{1!\,3!} = \frac{4\cdot3\cdot2\cdot1}{(1)(3\cdot2\cdot1)} = 4$$

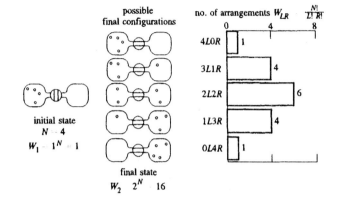

These numbers allow us to
construct a diagram analogous to Figure 11.7, shown above. Note that the number of
configurations is $N+1 = 5$, and the sum of the arrangements for all possible
configurations is W_2. Using Equation 11.15 we calculate

$$\Delta S = k_B\ln\frac{W_2}{W_1} = (1.38\times10^{-23}\,\text{J/K})\ln\frac{16}{1} = 3.83\times10^{-23}\,\text{J/K}\ \ (9.15\times10^{-24}\,\text{cal/K})$$

When $W_2 > W_1$, $\Delta S > 0$, reflecting a spontaneous process driven by the greater number
of arrangements, and hence likelihood, of the expanded state.

21. The total number of *configurations* is $(N+1)(N+2)/2 = 36$ for $N = 7$ molecules
distributed amongst 3 levels ε_0, ε_1, and ε_2; the total number of *arrangements* is $W = 3^N$
$= 2187$. We clearly don't want to write arrangements until we select the configurations
that have the right number of quanta. In an obvious notation, the possible configurations
are

$$0^71^02^0$$
$$0^61^12^0\quad 0^61^02^1$$
$$0^51^22^0\quad 0^51^12^1\quad 0^51^02^2$$
$$0^41^32^0\quad 0^41^22^1\quad 0^41^12^2\quad 0^41^02^3$$
$$0^31^42^0\quad 0^31^32^1\quad 0^31^22^2\quad 0^31^12^3\quad 0^31^02^4$$
$$0^21^52^0\quad 0^21^42^1\quad 0^21^32^2\quad 0^21^22^3\quad 0^21^12^4\quad 0^21^02^5$$
$$0^11^62^0\quad 0^11^52^1\quad 0^11^42^2\quad 0^11^32^3\quad 0^11^22^4\quad 0^11^12^5\quad 0^11^02^6$$
$$0^01^72^0\quad 0^01^62^1\quad 0^01^52^2\quad 0^01^42^3\quad 0^01^32^4\quad 0^01^22^5\quad 0^01^12^6\quad 0^01^02^7$$

Note that this configuration list cannot apply if the particles are electrons, since no more
than two of them could occupy a single nondegenerate level. Given that the levels are
equally spaced, the constraint of four quanta means that $N_1 + 2N_2 = 4$, which is
satisfied only for $(N_1,N_2) = (4,0)$, $(2,1)$, and $(0,2)$, or the configurations $0^31^42^0$, $0^41^22^1$,
and $0^51^02^2$. The combinatorial formula of the previous exercise yields
$$W_{340} = 7!/(3!4!0!) = 35, \quad W_{421} = 105, \quad \text{and} \quad W_{502} = 21$$
The explicit arrangements for each configuration would be constructed as in the previous

exercise; to save space we will not illustrate these here. The most probable configuration is clearly $0^4 1^2 2^1$; a plot of N_i *versus* ε is given at the right; the dashed curve is the predicted Boltzmann energy distribution. Note that other configurations, such as $0^3 1^2 2^2$, have larger Ws, but they do not satisfy the energy constraint. The decaying population with energy is characteristic of the Boltzmann distribution, although in Appendix D it is derived as a limiting case where all the N_i are $\gg 1$.

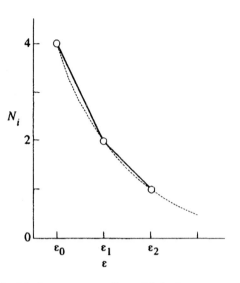

23. With a few exceptions such as that given in Example 11.1, a process for which Δn_{gas} $>(<)$ 0 leads to $\Delta S^\circ >(<)$ 0. This rule gives

a. $\Delta n_{gas} = -1$, $\Delta S^\circ < 0$

b. $\Delta n_{gas} = +1$, $\Delta S^\circ > 0$ (see, however, Example 11.1)

c. $\Delta n_{gas} = -2$, $\Delta S^\circ < 0$

d. $\Delta n_{gas} = -1$, $\Delta S^\circ < 0$

e. $\Delta n_{gas} = 0$, but we predict $\Delta S^\circ > 0$ based on configurational entropy. If we were to label the atoms, there is only one "arrangement" of the atoms possible on the reactant side, $H_a H_b$ and $Br_a Br_b$, but four possible on the product side, $H_a Br_a$, $H_a Br_b$, $H_b Br_a$, and $H_b Br_b$.

25. At constant pressure, Equation 11.20 can be converted to Equation 11.21 as follows:

$$\Delta S \geq \frac{q_P}{T}$$

$$\Delta S \geq \frac{\Delta H}{T}$$

$$T\Delta S \geq \Delta H$$

$$T\Delta S - \Delta H \geq 0$$

$$\Delta H - T\Delta S \leq 0$$

where the steps were ordered as prescribed below Equation 11.20, and allow us to write Equation 11.23, $\Delta G \leq 0$, where the free energy G is defined as in Equation 11.22, $G = H - TS$. Note that Equations 11.21 and 11.23 hold only at constant T and P. Nonetheless, G being a state variable, ΔG must be independent of path, and $\Delta G_{cycle} = 0$. Therefore a form of Hess's Law must hold that relates ΔGs for various reactions, allows the definition of a free energy of formation ΔG_f, and the calculation of ΔG° for a reaction from a formula analogous to Equation 10.30 for ΔH° and Equation 11.19 for ΔS°,

$$\Delta G^\circ = \sum_P \nu_P \Delta G_f^\circ(P) - \sum_R \nu_R \Delta G_f^\circ(R)$$

This will be given in Chapter 12 as Equation 12.2.

27. For an isothermal process in an ideal gas $\Delta H = 0$. Thus
$$\Delta G = -T\Delta S = -T[nR\ln(P_1/P_2)] = nRT\ln(P_2/P_1)$$
$$= -(298\ \text{K})(-0.890\ \text{cal/K}) = +265\ \text{cal}\ (+1110\ \text{J})$$

The increase in G is construed as an increased capacity to do work; it also tells us that isothermal compressions are never spontaneous.

29. Equations 10.30, 11.19, and 11.24 are required. In the usual manner, we write
$$C(s) + O_2(g) \rightarrow CO_2(g)$$
$$\Delta H^\circ = (-1)(0) + (-1)(0) + (1)(-94.052) = -94.052\ \text{kcal/mol}$$
$$\Delta S^\circ = (-1)(1.372) + (-1)(49.003) + (1)(51.06) = +0.69\ \text{cal/K mol}$$

Then

$$\Delta G^\circ = \Delta H^\circ - T\Delta S^\circ$$
$$= -94.052\ \text{kcal/mol} - \frac{(298\ \text{K})(0.69\ \text{cal/K mol})}{1000\ \text{cal/kcal}}$$
$$= -94.052 - 0.21 = -94.26\ \text{kcal/mol}$$

Since $\Delta G^\circ < 0$, the reaction is spontaneous. Note that, unlike the physical changes considered up to now, this chemical change owes all but 0.2% of its spontaneity to ΔH°, the small positive entropy change (note that $\Delta n_{\text{gas}} = 0$) being a feeble contributor here. This is typical for highly exothermic reactions; ΔH° dominates even when $\Delta n_{\text{gas}} \neq 0$.

31. The all-gas combustion of H_2 can be obtained by adding twice the vaporization of $H_2O(l)$ to the standard combustion (see Exercise 10.23 for the analogous methane combustion), with corresponding operations on the component ΔH°s and ΔS°s:

	ΔH°(kcal/mol)	ΔS°(cal/K mol)
$2H_2(g) + O_2(g) \rightarrow 2H_2O(l)$	-136.63	-78.00
$2H_2O(l) \rightarrow 2H_2O(g)$	$+21.038$	$+56.79$
$2H_2(g) + O_2(g) \rightarrow 2H_2O(g)$	-115.59	-21.21

We conclude that a turnaround temperature exists, given by
$$T_{\text{ta}} = \Delta H^\circ/\Delta S^\circ = (-115.59\ \text{kcal})/(-0.02121\ \text{kcal/K}) = 5452\ \text{K}.$$

This number is close to the sun's blackbody temperature (see Exercise 2.5); above this T water vapor would not only decompose, it would atomize, as discussed in Section 11.6 and in the next exercise. See Exercise 33 for a $\Delta G^\circ(T)$ graph of this reaction.

33. For the combustion of propane to water vapor, we again carry through our "foolproof" procedure using Equations 10.30, 11.19, and Appendix C,
$$C_3H_8(g) + 5O_2(g) \rightarrow 3CO_2(g) + 4H_2O(g).$$
$$\Delta H^\circ = -1(-24.821) - 5(0) + 3(-94.052) + 4(-57.796) = -488.52\ \text{kcal/mol}$$
$$\Delta S^\circ = -1(64.51) - 5(49.003) + 3(51.06) + 4(45.104) = +24.07\ \text{cal/K mol}$$

From Equation 11.24
$$\Delta G^\circ_{298} = -488.53\ \text{kcal/mol} - (298.15\ \text{K})(0.02407\ \text{kcal/K mol}) = -495.69\ \text{kcal/mol}$$

With $\Delta G^\circ_{298} \ll 0$, the reaction is overwhelmingly (explosively) spontaneous at 298 K; since it is favored by both H and S, there can be no turn-around temperature. Thus a graph of ΔG° versus T never crosses zero, as illustrated below.

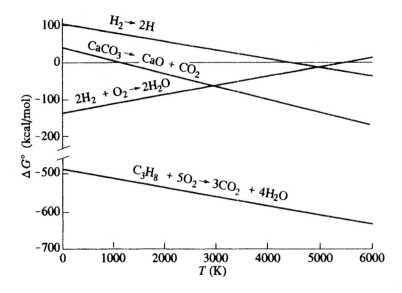

12

Free Energy
and Chemical Equilibrium

Your Chapter 12 GOALS:

- Determine ΔG°_{298} from $\Delta G^{\circ}_{f,298}$ values
- Calculate ΔG at non-standard conditions with the use of the reaction quotient Q
- Relate ΔG° to the equilibrium constant K
- Master the use of extent of reaction analysis to determine pressures, concentrations, or other factors in a reaction
- Use LeChâtelier's Principle to predict the response of a chemical equilibrium to changes in reaction conditions
- Express and perform calculations with the van't Hoff Equations, Equations 12.15 and 12.16
- Recognize the all-important role of $H_2O(l)$, $H^+(aq)$, and $OH^-(aq)$ in acid-base chemistry
- Calculate equilibrium constants from ΔG° and Q_{eq}
- Calculate pH's through an extent of reaction analysis
- Compare and contrast Equations 12.26, 12.27, and 12.28 and use them to calculate pH's
- Determine K values for the various Brønsted-Lowry acid-base reactions such as a weak base with a strong acid
- When necessary, utilize the common-ion effect to determine pH's
- Calculate pH's for monoprotic, diprotic, and triprotic acids using the appropriate approximations when necessary
- Sketch titration curves based on the characteristic points
- Explain the color changes of indicators
- Write equilibrium expressions for complexation reactions
- Determine the reaction quotient for multiphase equilibria and its implications for LeChâtelier's Principle
- Develop a deeper understanding of solubility that enhances and expands the general suppositions and conclusions that are stated as the salt solubility rules of Section 5.7
- Write ion product and solubility product expressions and perform calculations with them
- Outline the separation of ions that can be achieved via the scheme in Figure 12.13
- Extend your knowledge about non-ideal behavior that was formed in Section 9.5 by defining activity coefficients, ionic strength, and fugacity

Chapter 12 KEY EQUATIONS:

- 12.2, 12.5, 12.6, 12.8, 12.9, 12.14, 12.15, 12.16, 12.17, 12.20, 12.21, 12.22, 12.26, 12.27, 12.28, 12.29, 12.34, 12.42, 12.51, 12.56, 12.58

Overview

Chapters 1 through 8 mostly dealt with atoms, molecules, and their structure. Chapters 9 through 11 sum up thermodynamics. (Should you stay in chemistry, you will find that physical chemistry courses are usually divided into these two areas: one covers quantum material, the other covers thermodynamic material.) Chapters 12 through 18 can be viewed as the application chapters of *University Chemistry* in that atomistic and molecular material is combined with thermodynamic material to explore chemistry in solution, electron transfer, and the other remaining topics in the textbook. The increasing cumulative nature of the material, therefore, increases the complexity of the topics.

Equation 11.23 indicates that spontaneous reactions have $\Delta G \leq 0$. One could then ask if there exists a *degree of spontaneity*. Are some reactions or processes more spontaneous than others, with all other variables such as pressure and temperature being equal? The answer, of course, is yes. The more negative ΔG becomes, the more spontaneous the reaction is. While it takes some thought, there is a fundamental difference between the spontaneity of a reaction and the time it takes for the reaction to occur. We don't want to overuse words in this manual, and fundamental has been used several times previously and again here. The term, however, is *the* term that we need. "Fundamental" properties or topics are those that form the foundations of chemistry: entropy, internal energy, temperature, and wave-particle duality to name a few. The difference that we briefly highlight here is between *spontaneity* and *reaction rate* (Chapter 15). When chemists use the term "spontaneous", what they are really saying is that $\Delta G \leq 0$ is obtained when Equation 12.2 is applied to a chemical reaction, indicating that an overall free energy decrease would occur if the reactants could be converted into the products. It says *nothing* about *how fast* this reaction would occur. The world famous example of this is the very spontaneous reaction $2H_2(g) + O_2(g) \rightarrow 2H_2O(l)$ for which $\Delta G°_{298} = -113.374$ kcal from Equation 12.2. Nonetheless, a mixture of $H_2(g)$ and $O_2(g)$ will exist for eons before you would ever get a drop of water. This point is stressed now because we want you to keep this in mind as you read Chapter 12 and solve the Exercises. This way, the material of Chapter 15 will come to you as no shock, and its relevance to reactions will be more clear. Equation 12.6 is the equation that one needs to determine this spontaneity, and it is very important that you understand and never forget what the ° symbol means. In gas phase reactions it means all products and reactants are present at 1 atm. (See Exercise 2.) In solution reactions it means that all products and reactants are present at 1 M concentration. The ° symbol has *nothing* to do with temperature, and it *does not* indicate that $T = 298$ K. If it did, the notation $\Delta G°_{298}$ would be foolish and redundant. If you are confused about this, study Figure 11.8 until you convince yourself that $\Delta G°$ *depends on temperature*. The condition where all products and reactants are present at 1 atm or 1 M is rarely encountered, however, and Equation 12.6 gives chemists the ability to account for this fact. Further, the fact that $\Delta G = 0$ at equilibrium allows us to derive one of the most important equations in all of science, Equation 12.8. Memorize this equation. The last subsection of Section 12.1 begins the first time in the textbook that we consider the *extent of reaction* variable x. All typical stoichiometry problems assume that all of the reactants are converted into products. This is certainly true for many reactions, but by no means all reactions. x allows us to determine just how far a reaction proceeds. If x nearly equals the initial pressures or concentrations of the reactants, the reaction can be considered to go to completion. If x is small compared to the initial pressures or concentrations of the reactants, the reaction proceeds only slightly. Exercises 1 through 8 deal with Section 12.1 material. Exercise 6 is the first extent of reaction variable Exercise and you should be able to express the equilibrium partial pressures of H_2S, H_2, and S_2 as $P_0 - 2x$, $2x$, and x, respectively. Careful observance of the stoichiometric coefficients is required for any extent of reaction Exercise. Exercise 7 shows that you must always be thinking when x is obtained from the quadratic equation. Only one of the roots will be *chemically* valid.

The position of an equilibrium depends on temperature (and pressure, too), and the van't Hoff equation of Section 12.2 quantifies this temperature dependence. It is important to master this equation now for the Chapter 12 material, but it also appears in disguise, if you will, as Equation 14.8. A common mistake is to leave the $\Delta H°$ value in kcal, followed by division by R in cal. Exercise 11 is a marathon of a problem, but it contains, of course, a wealth of information. The qualitative dependence is to assume that exothermic reactions are shifted to the left by temperature increases and to the right for temperature decreases. Vice versa for endothermic reactions. We assume that we can apply this qualitative rule in all gas phase and solution phase reactions.

If you hear applause when solving the third parts of Exercises 18(a) and (b), it is us applauding our own trickery. You will not get the correct answer by blindly using Equation 12.21. These Exercises are designed to emphasize that ignoring the H^+ or OH^- concentration from the autoionization of water is only justified if the acid or base concentration is greater than $\sim 10^{-6}$. Here, the acid and base concentrations are extremely small and $[H^+] \approx 1 \times 10^{-7}$ is appropriate for each case. Exercise 21 reminds you that Equation 12.19 is endothermic and that higher temperatures shift the equilibrium to the right. Exercises 22 and 23 require Equation 12.8, and Exercises 24 and 25 require extent of reaction analysis, followed by setting up the equilibrium expression. To solve Exercise 32, identify the conjugate acid or base that is present; CN^-, $CH_3CH_2COO^-$, and OCl^- are conjugate bases of weak acids, and NH_4 is the conjugate acid of NH_3. Be sure to account for the fact that there are two $CH_3CH_2COO^-$ ions for every unit of $(CH_3CH_2COO)_2Ca$. For Exercise 42 note for formic acid that the log term of Equation 12.42 must equal 0.25 to obtain the desired pH since formic acid has $pK_a = 3.75$. From there you must determine what the ratio of moles for HA and A^- is and develop two equations in two unknowns V_A and V_B. One of these equations is $V_A + V_B = 0.500$ L.

The polyprotic acid Exercises again remind you to always be aware of the approximations that you are making. As long as $K_{a1} \gg K_{a2}$, the pH of a diprotic or polyprotic acid solution can be determined solely from the first ionization. This (very good) approximation greatly simplifies the math, yet it gives satisfactory results. In Exercise 46 note that shorthand notation is used for the oxalate ion, the structural formula for which is $^-OOCCOO^-$. Since oxalic acid (Table 12.3) is diprotic, follow the examination that was given for Equation 12.45 in the text. The titration Exercises are relatively simple for the start and midpoint pH values. The difficult part is calculating the pH at the endpoint(s). To do this, you must determine what species are present at equilibrium and, from that, determine which is responsible for the pH and what its equilibrium is. For example, in Exercise 50 at the endpoint all of the formic acid has been stoichiometrically converted to $HCOO^-$ by the NaOH. Since $HCOO^-$ is the conjugate base of the acid, the solution will be basic at the endpoint due to the reaction $HCOO^- + H_2O \rightleftharpoons HCOOH + OH^-$. For your sketches to be accurate, do not ignore the flat, buffer regions which are essential facets of weak acid or base titration curves. For Exercise 56, assume complete conversion to products, which means that $[Ni(NH_3)_6^{2+}] \approx [NiCl_2]_0$, and use the approximate expressions derived in Equation 12.51.

Because $CO_2(g)$ is the only pure, non-condensed phase present, K directly equals the CO_2 partial pressure at equilibrium in Exercise 60. Further, you must determine $\Delta G°_{1473}$ and then use Equation 12.8. The approach is similar for Exercises 61, 62, and 63 with two of them requiring an extent of reaction analysis. Note that Exercise 64 verifies the given $[H_2S]_0$ in Exercise 45. Solubility and precipitation exercises, such as Exercises 66, 68, 69, and 70, almost always involve the comparison of Q_{ion} and K_{sp}. While Q_{ion} might have the "ion" subscript, its role in precipitation matters is the same as is it in Equation 12.6; it helps to determine in which direction a reaction will proceed. Exercise 72 can be solved by determining $\Delta H°$ from Appendix C, followed by use of Equation 12.15. This also gives us the chance to repeat that Equation 12.15 is just Equation 11.24 expressed in terms of K, and Exercise 72 can also be solved by the more circuitous route of calculating $\Delta H°_{298}$, $\Delta S°_{298}$, and $\Delta G°_{368}$ from them, and then using Equation 12.8. For Exercise 74 begin by considering the reverse reaction.

The non-ideal corrections are a difficult concept by themselves. You may, in fact, not have fully understood the van der Waals correction in Chapter 9. What is worse is that they put a dent in the security you have in your knowledge. Now you know that, in addition to molarity, you must consider activities and ionic strength where appropriate and that, in addition to pressure, you must worry about fugacity. Attempt to understand this material if it is assigned to you, but do not dwell on fugacity, activity, or ionic strength. Advanced courses cover this in much greater detail, and it is presented to you now to allow you to become familiar with the terms, if nothing else, or, perhaps, the mere concept that the ideal equations we often use are not exactly right. Note, however, that Exercise 77 illustrates again that ideal behavior is approached when the molecules don't "know" if the other molecules are there. (Such an assumption is made in Dalton's Law of Partial Pressures.) If the intermolecular forces (attractive or repulsive) are not able to act, no deviations from ideality will occur.

12 Free Energy and Chemical Equilibrium

1. Following the steps shown in Equation 12.5,

$$\Delta G = 3\Delta G_f(O_2) - 2\Delta G_f(O_3)$$

$$= 3\Delta G_f^{\circ}(O_2) + 3RT\ln P_{O_2} - 2\Delta G_f^{\circ}(O_2) - 2RT\ln P_{O_3}$$

$$= \left[3\Delta G_f^{\circ}(O_2) - 2\Delta G_f^{\circ}(O_3)\right] + \left[RT\ln P_{O_2}^3 - RT\ln P_{O_3}^2\right]$$

$$= \Delta G^{\circ} + RT\ln\frac{P_{O_2}^3}{P_{O_3}^2}$$

Note that the coefficients become the exponents in the pressure ratio. ΔG_{298}° from Appendix C is -78.0 kcal/mol; we also use $RT = (1.9872$ cal/K mol$)(298.15$K$) = 592.5$ cal/mol. The above formula then gives

$$\Delta G_{298} = -78.0 + 0.5925\ln\frac{(0.200)^3}{(0.00100)^2} = -72.7 \text{ kcal/mol } (-304 \text{ kJ/mol})$$

Since $\Delta G < 0$, the reaction is spontaneous as written under these conditions. From this example we generalize that, for "large" $|\Delta G^{\circ}|$—roughly 20 kcal/mol or greater in magnitude—the presence of comparable amounts of reagents and products will not alter the spontaneous direction of the reaction as determined by the sign of ΔG°.

3. Equation 12.6 gives

$$Q = \frac{P_{SO_3}^2}{P_{SO_2}^2 P_{O_2}} \quad \text{and} \quad \Delta G = \Delta G^{\circ} + RT\ln\frac{P_{SO_3}^2}{P_{SO_2}^2 P_{O_2}}$$

From Appendix C and Equation 12.2, $\Delta G_{298}^{\circ} = -33.88$ kcal $(-141.8$ kJ$)$. K follows from Equation 12.8:

$$K = e^{-\Delta G^{\circ}/RT} = \exp\left(-\frac{-33.88 \text{ kcal/mol}}{0.5925 \text{ kcal/mol}}\right) = 6.82 \times 10^{24}$$

According to Equation 12.9, if $Q < K$, $\Delta G < 0$, the reaction will proceed to more products. We can therefore predict whether the reaction will be spontaneous by calculating Q and comparing it to K. From the above expression

$$Q = (0.25)^2/[(0.50)^2(0.10)] = 2.5$$

Thus, $Q \ll K$, and the reaction is spontaneous in the forward direction. Since we already have K, we can use Equation 12.9 to find

$$\Delta G = (0.5925 \text{ kcal/mol})\ln(2.5 / 6.82\times 10^{24}) = -33.3 \text{ kcal/mol } (-139 \text{ kJ/mol})$$

The sign is negative, indicating a spontaneous reaction. Recall our observation in the solution to Exercise 1.

5. Keeping in mind that $K = Q_{eq}$, Equation 12.6 gives

$$K = \frac{P_{NH_3}^2}{P_{N_2} P_{H_2}^3} = \frac{(1.74)^2}{(7.07)(21.20)^3} = 4.49 \times 10^{-5}$$

From Equation 12.8

$$\Delta G_{723}^{\circ} = -RT\ln K = -(0.0019872 \text{ kcal/K mol})(723 \text{ K})\ln(4.49\times 10^{-5})$$

$$= +14.4 \text{ kcal/mol} \ (60.2 \text{ kJ/mol})$$

7. An extent-of-reaction analysis yields

	$Cl_2(g)$	$+$	$Br_2(g)$	$\rightleftharpoons$	$2BrCl(g)$
initial	P_0		P_0		0
equilibrium	$P_0 - x$		$P_0 - x$		$2x$

From Appendix C $\Delta G^{\circ}_{298} = -0.46$ kcal. Next, K is obtained from Equation 12.8:

$$K = e^{-\Delta G^{\circ}/RT} = \exp\left(-\frac{-0.46 \text{ kcal/mol}}{0.5925 \text{ kcal/mol}}\right) = 2.17$$

The equilibrium reaction quotient is

$$K = \frac{P^2_{BrCl}}{P_{Cl_2} P_{Br_2}} = \frac{(2x)^2}{(P_0 - x)^2}$$

This is a perfect square, so we don't need to use the quadratic formula. Taking the positive square root of both sides and solving for x we find

$$x = \frac{P_0}{1 + 2/\sqrt{K}} = \frac{200.}{1 + 2/\sqrt{2.17}} = 84.8 \text{ torr}$$

Note that, because $\Delta n_{gas} = 0$, K is unitless, and we were able to use torr directly in calculating x. Lastly, the expressions from our table give the equilibrium partial pressures as

$$P_{Cl_2} = P_{Br_2} = P_0 - x = 200. - 84.8 = 115 \text{ torr} \qquad P_{BrCl} = 2x = 170. \text{ torr}$$

Since the equilibrium was linear in x in this case, there was no chance of selecting the wrong root; if the starting pressures were not equal, however, we would have to solve a quadratic. The value of x is verified by back-substitution into the expression for K:
$$[2(84.8)]^2/(200. - 84.8)^2 = 2.17.$$

9. For the effect of temperature we calculate ΔH° from Appendix C for each of the reactions, while for the pressure effect we need Δn_{gas}. LeChâtelier's Principle predicts that the equilibrium position will shift to the right (forward reaction favored) upon increases in T and P when $\Delta H^{\circ} < 0$ and $\Delta n_{gas} < 0$, respectively, and to the left when $\Delta H^{\circ} > 0$ and $\Delta n_{gas} > 0$.

a. $\Delta H^{\circ} = +25.9$ kcal ($+108$ kJ) $\Rightarrow$ shift to right; $\Delta n_{gas} = +1 \Rightarrow$ shift to left.

b. $\Delta H^{\circ} = -99.5$ kcal (-416 kJ) $\Rightarrow$ shift to left; $\Delta n_{gas} = -2 \Rightarrow$ shift to right.

c. $\Delta H^{\circ} = -34.3$ kcal (-143 kJ) $\Rightarrow$ shift to left; $\Delta n_{gas} = +1 \Rightarrow$ shift to left.

d. $\Delta H^{\circ} = -23.5$ kcal (-98 kJ) $\Rightarrow$ shift to left; $\Delta n_{gas} = 0 \Rightarrow$ no effect.

e. $\Delta H^{\circ} = +119$ kcal ($+498$ kJ) $\Rightarrow$ shift to right; $\Delta n_{gas} = +6 \Rightarrow$ shift to left.

11. a. A bond-breaking and therefore endothermic process that creates more gas will be favored by an increase in temperature but not by an increase in pressure. The predicted temperature effect is already borne out by the increasing K values given.

b. The extent-of-reaction analysis gives

	$Cl_2 \rightleftharpoons 2Cl$		Total pressure P
initial	P_0	0	P_0
equilibrium	$P_0 - x$	$2x$	$P_0 + x$

As in Example 12.2, the partial pressures can be re-expressed in terms of the total pressure P, giving

$$K = \frac{P_{Cl}^2}{P_{Cl_2}} = \frac{(2x)^2}{P - 2x}$$

$$x^2 + \tfrac{1}{2}Kx - \tfrac{1}{4}KP = 0$$

$$P_{Cl} = 2x = -\tfrac{1}{2}K + \sqrt{\tfrac{1}{4}K^2 + KP}$$

The positive root will result from choosing the positive sign. The mole fraction $X_{Cl} = P_{Cl}/P = 2x/P$ can then be written as

$$X_{Cl} = \frac{K}{2P}\left(\sqrt{1 + \frac{4P}{K}} - 1\right)$$

$$\rightarrow \sqrt{\frac{K}{P}}, \quad P \gg K$$

As predicted, increasing P reduces X_{Cl}, as seen clearly using the high-P limit. (We can derive an analogous low-P limit using the binomial expansion of the square root, which yields $P_{Cl}/P \rightarrow 1 - P/K$, also decreasing with increasing P.)

c. Applying the results of part (b) for K = 0.0129 and P = 1.00 atm yields

$$X_{Cl} = 0.107, \quad P_{Cl} = X_{Cl}P = 0.107 \text{ atm}$$
$$P_{Cl_2} = P - P_{Cl} = 1.00 - 0.107 = 0.89_3 \text{ atm}$$

while for P = 10.00 atm

$$X_{Cl} = 0.0353 \quad P_{Cl} = 0.353 \text{ atm} \quad P_{Cl_2} = 9.64_7 \text{ atm}$$

$\tfrac{1}{2}X_{Cl}$ is also the approximate fractional dissociation α; the exact fraction is
$$\alpha = \tfrac{1}{2}P_{Cl}/P_0 = x/(P - x) = \tfrac{1}{2}X_{Cl}/(1 - \tfrac{1}{2}X_{Cl})$$
This yields $\alpha = 0.0565$ at 1.00 atm and 0.0180 at 10.00 atm. Thus, as predicted in part (a), the extent of dissociation is lower at higher pressure.

d. Solving Equation 12.15 for ΔH° gives

$$\Delta H^\circ = -\left(\frac{1}{T_2} - \frac{1}{T_1}\right)^{-1} R \ln\frac{K_2}{K_1}$$

$$= -\left(\frac{1}{1600 \text{ K}} - \frac{1}{1200 \text{ K}}\right)^{-1} \frac{0.0019872 \text{ kcal}}{\text{K mol}} \ln\frac{0.0129}{2.48 \times 10^{-5}}$$

$$= +59.7 \text{ kcal/mol} \ (250. \text{ kJ/mol})$$

endothermic as predicted. From Appendix C $\Delta H^\circ_{298} = 58.164$ kcal (243.36 kJ). The small but significant discrepancy may indicate that ΔH° is temperature dependent.

13. From Appendix C, $\Delta H^\circ = -22.04$ kcal (-92.22 kJ), while $\Delta n_{gas} = -2$; thus, products are favored at low T and high P. Industrially, P is high, but T is high as well. High T increases the reaction rate, as we will discuss in Chapter 15, and the chosen temperature represents a compromise between complete reaction and rapid reaction. There are two equivalent ways to find K_{700}: find ΔG°_{298} from Appendix C, K_{298} from Eq. 12.8, and K_{700} from the van't Hoff Equation 12.15; or, as we prefer, find ΔS°_{298} from Appendix C, ΔG°_{700} from Equation 11.24, and K_{700} from Eq. 12.8. From Appendix C $\Delta S^\circ = -47.45$ cal/K, and

$$\Delta G^\circ_{700} = -22.04 - (700)(-0.04745) = +11.18 \text{ kcal}$$
$$K_{700} = \exp[-(11,180)/(1.9872)(700)] = 3.23 \times 10^{-4}$$

This small K yields small amounts of product ($\sim 20\%$ conversion), but the exhaust gases are cooled to trap liquid NH_3 and then recycled, leading eventually to 85-90% conversion.

15. From Appendix C $\Delta H^\circ = 22.071$ kcal (92.345 kJ), and $\Delta S^\circ = 33.2$ cal/K (139 J/K). Equation 11.24 gives $\Delta G^\circ_{773} = -3.593$ kcal, and Equation 12.8 yields $K = 10.4$. Previous Exercises show the use of Equation 12.8. For the cracking fraction, we do an extent-of-reaction analysis:

				Total pressure
	C_4H_{10} $\rightleftharpoons$	C_2H_6 +	C_2H_4	P
initial	P_0	0	0	P_0
equilibrium	$P_0 - x$	x	x	$P_0 + x$

As in Example 12.2, we can express the reaction quotient in terms of P, and solve the resulting quadratic to obtain

$$x = K\{[1 + P/K]^{1/2} - 1\} = 10.4\{[1 + 1.00/10.4]^{1/2} - 1\} = 0.489 \text{ atm}$$

The cracking fraction is $x/(P-x) = 0.957$. Since $\Delta H^\circ > 0$ and $\Delta n_{gas} = +1$, cooling the reaction will reduce the yield, as will increasing P.

17. On a per mole basis, Equation 12.17 gives

$$\Delta G = RT\ln\frac{c_2}{c_1} = (0.5925 \text{ kcal/mol})\ln\frac{0.025}{1.00} = -2.19 \text{ kcal/mol} \quad (-9.14 \text{ kJ/mol})$$

Dilution is thus spontaneous, and the process of diluting a solution reduces its free energy, making it less potent.

19. a. Inverting the definition of pH yields $[H^+] = 10^{-pH}$ and similarly for pOH = 14.00 − pH. Thus $[H^+] = 10^{-10.68} = 2.09 \times 10^{-11}$ M and $[OH^-] = 10^{-3.32} = 4.79 \times 10^{-4}$ M. Parts (b) through (f) proceed similarly; see text answer key.

21. For the autoionization of water, Equation 12.19, $\Delta H^\circ_{298} = 13.345$ kcal/mol. K for the autoionization at 298 K is 1.0×10^{-14}, and it is given the symbol K_w. The van't Hoff equation shows that K_w depends on temperature. From the reaction's endothermicity and

LeChâtelier's Principle we expect $K_w < 1.0 \times 10^{-14}$ at 273 K and $K_w > 1.0 \times 10^{-14}$ at 310 K and 373 K. In Equation 12.15, K_1 will be K_w at 298 K. Expressing Equation 12.15 in a general form for this problem gives

$$\ln \frac{K_w}{1.0 \times 10^{-14}} = -\frac{13{,}345 \text{ cal/mol}}{1.9872 \text{ cal/K mol}} \left[\frac{1}{T} - \frac{1}{298.15\text{K}} \right]$$

Then $K_w = 1.26 \times 10^{-15}$, 2.37×10^{-14}, and 9.18×10^{-13} for 273 K, 310 K, and 373 K, respectively, giving pK_ws of 14.90, 13.63, and 12.04. Our qualitative K_w predictions were correct. Then pH $= \frac{1}{2} pK_w = 7.45$, 6.81, and 6.02. N.B. Most chemistry students would quickly answer the question, "What is the pH of a neutral aqueous solution?" From the above, we have shown that the answer "7.00" is only true at 298 K.

23. Equation 12.8 gives the range of ΔG°_{298} from $-0.592 \ln(10^{-4}) = 5.5$ kcal/mol (23 kJ/mol) to $-0.592 \ln(10^{-10}) = 13.6$ kcal/mol (57 kJ/mol). This relation also gives us a rough rule-of-thumb: each order-of-magnitude (factor of 10) change in K corresponds to an increment of $(0.6)(2.3) = 1.4$ kcal/mol in ΔG°_{298}. This rule gives us $\Delta G^{\circ} = 5.6$ to 14 kcal/mol, as found by direct calculation.

25. The pH yields $[H^+] = 10^{-2.09} = 0.00813$ M $= x$. Equation 12.25 then gives

$$K_a = \frac{x^2}{c - x}$$
$$= \frac{(0.00813)^2}{0.150 - 0.008} = 4.66 \times 10^{-4}$$

and $pK_a = -\log K_a = 3.33$. The approximation of Equation 12.26 yields $pK_a = 2pH - pc = 4.18 - 0.82 = 3.36$.

27. The Lewis structures are:

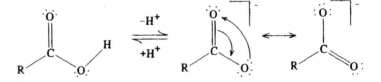

The carbon is sp^2 hybridized (120° angles), creating σ overlap with the oxygen $p\sigma$ orbitals, and leaving one unhybridized p orbital to engage in resonance (VB) or delocalized π bonding (MO) with the oxygen $p\pi$s once the anion forms. This stabilizes the anion, and helps to interpret the thermoneutrality of the ionization process. The negative entropy change reflects increased organization of the solvent (water) engendered by the ions.

29. The equilibrium, charge balance, and mole balance conditions are:
    1. base equilibrium $K_b = [NH_4^+][OH^-] / [NH_3]$
    2. water equilibrium $K_w = [H^+][OH^-]$
    3. charge balance $[NH_4^+] + [H^+] = [OH^-]$
    4. mole balance $[NH_3]_0 = [NH_3] + [NH_4^+]$
Note that $[H_2O]$ does not appear in either the base reaction quotient or water ionization.

The base equilibrium condition (1) is the foundation for incorporating the effect of water ionization. Letting $x = [OH^-]$ and $c = [NH_3]_0$, we express $[NH_4^+]$ and $[NH_3]$ in terms of x, c, and K_w by carrying out the steps

From (2) $\qquad\qquad [H^+] = K_w/[OH^-] = K_w/x$

From (3) $\qquad\qquad [NH_4^+] = [OH^-] - [H^+] = x - K_w/x$

From (4) $\qquad\qquad [NH_3] = [NH_3]_0 - [NH_4^+] = c - x + K_w/x$

Substituting these expressions into (1) yields

$$K_b = \frac{x\left(x - \dfrac{K_w}{x}\right)}{c - x + \dfrac{K_w}{x}}$$

This is identical in form to Equation 12.30 with a suitable change in definition of variables. The identical form arises because the stoichiometry of ionization is isomorphic with the weak acid case, and water ionization is symmetrical in H^+ and OH^-. This reduces to the simpler Equation 12.31 if $[H+] \ll [OH^-]$, that is, for pH > 8 or so.

31. For the following: S = strong; W = weak; A = acid; B = base; C = conjugate. The key to finding K for these reactions is recognizing the Brønsted-Lowry combinations WA+SB, SA+WB, and WA+WB, for which $K = K_a/K_w$, K_b/K_w, and $K_a K_b/K_w$, respectively. The component Ks come from Tables 12.1 and 12.2; ΔG° is then obtained from Equation 12.8.

a. $\quad CH_3NH_2 + HCl \rightleftharpoons CH_3NH_3^+ + Cl^-$
    $\qquad$ WB $\qquad$ SA $\qquad\qquad$ CA $\qquad$ CB
    $K = K_b/K_w = 4.4\times10^{-4}/1.00\times10^{-14} = 4.4\times10^{10}$
    $\Delta G^\circ = -RT\ln K = -0.592\ln 4.4\times10^{10} = -14.5$ kcal/mol $= -60.8$ kJ/mol

b. $\quad HNO_2 + KOH \rightleftharpoons KNO_2 + H_2O$
    $\qquad$ WA $\qquad$ SB $\qquad\quad$ CB $\qquad$ CA
    $K = K_a/K_w = 4.6\times10^{-4}/1.00\times10^{-14} = 4.6\times10^{10}$
    $\Delta G^\circ = -0.592\ln 4.6\times10^{10} = -14.5$ kcal/mol $= -60.9$ kJ/mol

c. $\quad NH_3 + HCOOH \rightleftharpoons NH_4^+ + HCOO^-$
    $\qquad$ WB $\qquad$ WA $\qquad\qquad$ CA $\qquad$ CB
    $K = K_a K_b/K_w = (1.77\times10^{-4})(1.79\times10^{-5})/1.00\times10^{-14} = 3.17\times10^5$
    $\Delta G^\circ = -0.592\ln 3.17\times10^5 = -7.50$ kcal/mol $= -31.4$ kJ/mol

d. $\quad 2NH_4Cl + Ba(OH)_2 \rightleftharpoons 2NH_3 + BaCl_2 + 2H_2O$
    $\qquad$ WA $\qquad\quad$ SB $\qquad\qquad$ CB $\qquad\qquad$ CA
    The net ionic equation here contains a common factor of 2, since $Ba(OH)_2$ supplies two OH^-; thus as written ΔG° is double that for a single proton transfer, and K is the square. Then, noting that here $K_a = K_w/K_b$, we have
    $K = (K_a/K_w)^2 = (1/K_b)^2 = (1/1.79\times10^{-5})^2 = 3.12\times10^9$
    $\Delta G^\circ = -0.592\ln 3.12\times10^9 = -13.0$ kcal/mol $= -54.2$ kJ/mol

e. $CH_3COONa + HIO_3 \rightleftharpoons CH_3COOH + NaIO_3$
       WB   WA   CA   CB

Noting that here $K_b = K_w/K_a'$,
$$K = K_a K_b / K_w = K_a/K_a' = 0.16/(1.76 \times 10^{-5}) = 9.1 \times 10^3$$
$$\Delta G^\circ = -0.592\ln 9.1 \times 10^3 = -5.4 \text{ kcal/mol} = -23 \text{ kJ/mol}$$

33. The new aspects of this system relative to the weak base equivalent of Equation 12.31 are the use of Equation 12.34 to obtain K_b and the inclusion of the Na^+ ion in the charge balance condition, where it appears as $[Na^+]_0$, not being involved in any equilibria. See text answer key.

35. The equilibrium condition that applies here is Equation 12.40, with $c_1 = [CH_3COOH]_0$, $c_2 = [HBr]_0$, and $x = [CH_3COO^-]$,
$$K_a = \frac{(c_2 + x)x}{c_1 - x} \approx \frac{c_2 x}{c_1}$$
where the approximation results from applying LeChâtelier's Principle as discussed in the text. Substituting the concentrations, including dilution after mixing, and K_a from Table 12.1,
$$x = [CH_3COO^-] = \frac{c_1 K_a}{c_2} = \frac{(0.125)(1.76 \times 10^{-5})}{0.025} = 8.8 \times 10^{-5} \text{ M}$$
and pH $\approx -\log c_2 = 1.60$. Note that $x \ll c_1, c_2$, justifying our approximation.

37. The unapproximated equilibrium condition of Equation 12.41 is readily shown to lead to the quadratic equation
$$x^2 + (K_a + [A^-]_0)x - K_a[HA]_0 = 0$$
Solving this equation is numerically well-conditioned when both $[HA]_0$ and $[A^-]_0$ are much larger than K_a. The solution to this equation reduces to the Henderson-Hasselbalch Equation 12.42 only when $[A^-]_0$ is much larger than both $K_a[HA]_0$ and K_a; both of these conditions are fulfilled for both cases considered in Example 12.7, and are also what is necessary for an effective buffer solution. Using suitably diluted concentrations, the positive root of the quadratic equation yields (a) pH $= 4.755$ and (b) pH $= 5.232$; these agree to two decimal places with the HH results quoted in the example. The ionization of water as a competing equilibrium has not been included in our treatment; this is unimportant when both $[HA]_0$ and $[A^-]_0$ are large.

39. a. The reaction is
   $$CH_2ClCOOH + NaOH \rightleftharpoons CH_2ClCOONa + H_2O$$
   This is a WA+SB neutralization reaction like Equation 12.32, and
   $$K = K_a/K_w = 1.40 \times 10^{-3}/1.0 \times 10^{-14} = 1.4 \times 10^{11}$$
   The large K assures a stochiometric titration.

   b. From the solution volumes and concentrations we have
   $$\text{mmol NaOH} = (0.200)(20.0) = 4.00 \text{ mmol}$$
   $$\text{mmol CH}_2\text{ClCOOH} = (0.500)(25.0) = 12.50 \text{ mmol}$$

With 1:1 stoichiometry, this implies that 4.00 mmol $CH_2ClCOONa$ have been formed, and $12.50 - 4.00 = 8.50$ mmol $CH_2ClCOOH$ remain. HH then yields

$$pH = 2.85 + \log\frac{4.00}{8.50} = 2.52$$

c. To make a 1:1 buffer, we need to get to the midpoint of the titration, which will correspond to equal concentrations and mmols of acid and salt. Thus we need to add $6.25 - 4.00 = 2.25$ mmol NaOH or $(2.25 \text{ mmol})/(0.200 \text{ mmol/mL}) = 11.25$ mL NaOH.

41. From Table 12.2, $pK_a = 14.00 - 4.75 = 9.25$ for NH_4^+. Together with the given concentrations, the buffer Equation 12.42 then gives us

$$pH = pK_a + \log\frac{[NH_3]_0}{[NH_4^+]_0}$$

a. $pH = 9.25 + \log\frac{0.25}{0.25} = 9.25$

b. $pH = 9.25 + \log\frac{0.10}{0.50} = 8.55$

We expect that adding NH_4Cl to $NH_3(aq)$ should drive the weak-base equilibrium to the left, thereby reducing $[OH^-]$ (increasing pOH) and increasing $[H^+]$ (decreasing pH). This is confirmed by comparing the above pH values to those obtained for NH_3 solutions of the same concentrations from the weak-base analogue of Equation 12.26, which yields (a) $pH = 14 - \frac{1}{2}(4.75 + 0.60) = 11.32_5$, and (b) $pH = 11.13$, both higher by more than 2 pH units than the above buffer values.

43. As mentioned in the polyprotic acid subsection, in H_3PO_3 one of the H atoms is bonded directly to P. Thus, it is a diprotic acid despite its resemblance to the triprotic H_3PO_4. From Table 12.3, $pK_{a1} = 2.00$ ($K_{a1} = 0.0100$) and $pK_{a2} = 6.59$ ($K_{a2} = 2.57\times10^{-7}$); thus H_3PO_3 resembles H_2CO_3, the text example, in the great disparity between its stepwise Ks. By analogy with the carbonic acid equations, Equations 12.43 and 12.44, the pH is controlled by the first ionization, allowing the use of the monoprotic WA formulas, Equation 12.27 or 12.28, to determine $[H^+]$. Using 12.28, we have

$$[H^+] \approx \sqrt{cK_{a1}} - K_{a1}/2$$
$$= [(0.200)(0.0100]^{1/2} - 0.0100/2 = 0.0397M; \quad pH = 1.40$$
$$[H_2PO_3^-] \approx [H^+] = 0.0397 \text{ M}$$
$$[HPO_3^{2-}] \approx K_{a2} = 2.57\times10^{-7} \text{ M}$$

The exact quadratic solution of step 1 ionization yields 0.0400 M H^+, while Equation 12.26 gives 0.045 M. Exact solution of the coupled equilibrium equations 12.43, including water ionization, charge balance, and mole balance, yields results indistinguishable from the above.

45. a. In view of the widely different K_as for H_2S from Table 12.3, the results found in the paragraph containing Equation 12.44 apply, namely, that the pH may be found from

the first ionization step only, and that $[S^{2-}] \approx K_{a2}$. The numbers are

$$pH = \tfrac{1}{2}(pK_{a1} + pc) = \tfrac{1}{2}(7.04 + 1.00) = 4.02$$
$$[HS^-] = [H^+] = 10^{-4.02} = 9.5 \times 10^{-5} \text{ M}$$
$$[H_2S] = K_{a2} = 10^{-13.89} = 1.3 \times 10^{-14} \text{ M}$$

b. When $[H^+]$ from a strong acid is added to yield a known pH, $[HS^-]$ is computed from Equation 12.40, and $[S^{2-}]$ is determined by the overall equilibrium Equation 12.49; we have

$$pHS^- = pK_{a1} + pc - pH = 7.04 + 1.00 - 3.00 = 5.04$$
$$pS^{2-} = pK_{a1} + pK_{a2} + pc - 2pH = 7.04 + 13.89 + 1.00 - 6.00 = 15.93$$

or $[HS^-] = 9.1 \times 10^{-6}$ M and $[S^{2-}] = 1.2 \times 10^{-16}$ M, two orders of magnitude smaller than part (a). Exact solution of the coupled equilibria yield the same concentrations to better than 3S.

47. With 100% ionization in the first step, this becomes a common ion problem like that shown in Equation 12.39, where bisulfate HSO_4^- becomes the weak acid and the step 1 H^+ serves as the common ion. Here $[H^+]_0 = [HSO_4^-]_0 = c$, and the equilibrium condition is written and solved as

$$K = \frac{(c + x)x}{c - x}$$
$$x^2 + (K + c)x - cK = 0$$
$$x = \tfrac{1}{2}(K+c)\left(\sqrt{1 + 4cK/(K+c)^2} - 1\right)$$
$$x = 0.5(0.062)([1 + 4(0.050)(0.012)/(0.062)^2]^{1/2} - 1) = 0.0085$$

Thus $[SO_4^{2-}] = x = 8.5 \times 10^{-3}$ M, $[HSO_4^-] = c - x = 0.0415$ M, $[H^+] = c + x = 0.0585$ M and pH $= 1.23$. Exact solution of the competing equilibria gives results indistinguishable from these; we find that $[H_2SO_4]$ is very small at equilibrium $(2.4 \times 10^{-5}$ M), supporting our assumption of complete step 1 ionization.

49. We expect the weak-base version of Equation 12.26 to apply, along with the base analogue of Equation 12.44. Thus
$$pOH \approx \tfrac{1}{2}(pK_{b1} + pc) = \tfrac{1}{2}(6.05 + 0.60) = 3.32; \quad pH = 10.68$$
$$[N_2H_6^{2+}] \approx K_{b2} = 1.0 \times 10^{-14}$$

51. Comparing the volumes reveals that 13.93 is half of 27.86. If 27.86 mL is the end point, then the midpoint is at 13.93 mL. As Equation 12.42 shows, this is where pH $=$ pK_a. Thus, $pK_a = 4.19$, and $K_a = 6.5 \times 10^{-5}$. Although with only this information we can't calculate the pH at any other points, we know we have a weak acid, and, under typical laboratory conditions, we can closely estimate the pH titration curve. In the graph on the next page we have chosen 0.1 M acid and base, with a 20 mL initial volume as in Figure 12.10. (Inspection of Table 12.1 suggests the acid is benzoic.)

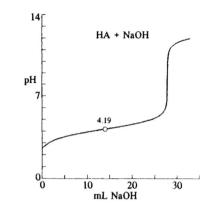

53. The neutralization reaction is $H_2SeO_3 + 2NaOH \rightarrow Na_2SeO_3 + 2H_2O$. From Table 12.3, $pK_{a1} = 2.46$ and $pK_{a2} = 7.31$. From stoichiometry at the two endpoints we have

End1: $[H_2SeO_3]_0 = (0.2000 \text{ M NaOH})(15.71 \text{ mL})/(10.00 \text{ mL}) = 0.3142 \text{ M}$

End2: $[H_2SeO_3]_0 = \frac{1}{2}(0.2000 \text{ M NaOH})(31.38 \text{ mL})/(10.00 \text{ mL}) = 0.3138 \text{ M}$

Taking the average, we obtain $[H_2SeO_3]_0 = 0.3140$ M. The widely different Ks imply that we can use the approximations surrounding Equations 12.44 to 12.48 to compute the five important points on the titration curve. We have

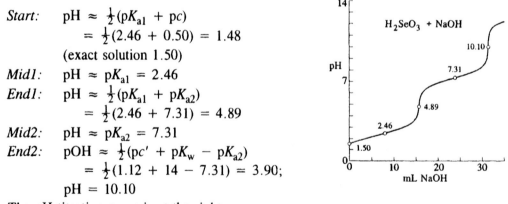

Start: $pH \approx \frac{1}{2}(pK_{a1} + pc)$
            $= \frac{1}{2}(2.46 + 0.50) = 1.48$
           (exact solution 1.50)

Mid1: $pH \approx pK_{a1} = 2.46$

End1: $pH \approx \frac{1}{2}(pK_{a1} + pK_{a2})$
            $= \frac{1}{2}(2.46 + 7.31) = 4.89$

Mid2: $pH \approx pK_{a2} = 7.31$

End2: $pOH \approx \frac{1}{2}(pc' + pK_w - pK_{a2})$
            $= \frac{1}{2}(1.12 + 14 - 7.31) = 3.90;$
           $pH = 10.10$

The pH titration curve is at the right.

55. a. The net ionic equation is $2H^+(aq) + CO_3^{2-}(aq) \rightarrow H_2CO_3(aq)$, assuming that the carbonic acid concentration is low enough that $CO_2(g)$ is not evolved. This reaction is the reverse of the complete ionization reaction Equation 12.49, and hence its equilibrium constant is the inverse of that given there:

$$K \approx \frac{1}{K_{a1}K_{a2}} = \frac{1}{(4.3 \times 10^{-7})(4.8 \times 10^{-11})} = 4.8 \times 10^{16}$$

The product is thus highly favored, and this titration will be stoichiometric.

b. We have mmol $Na_2CO_3 = 171.4 \text{ mg} / 105.99 \text{ mg/mmol} = 1.617$ mmol, implying

$$[CO_3^{2-}]_0 = c = 1.617 \text{ mmol} / 50.00 \text{ mL} = 0.03234 \text{ M}$$

The 2:1 H^+ to CO_3^{2-} stoichiometry then yields

$$[HCl] = 1.617 \text{ mmol Na}_2CO_3 \left(\frac{2 \text{ mmol HCl}}{1 \text{ mmol Na}_2CO_3} \right) / (16.07 \text{ mL}) = 0.2012 \text{ M}$$

At the equivalence point, then,

$$[H_2CO_3]_0 = c' = 1.617 \text{ mmol} / (50.00 + 16.07 \text{ mL}) = 0.02448 \text{ M}$$

c. We use approximations equivalent to those for the diprotic acid case Equations 12.44 to 12.48.

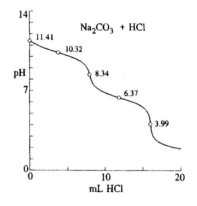

Start: $pOH \approx \frac{1}{2}(pK_w - pK_{a2} + pc)$
$= \frac{1}{2}(14.00 - 10.32 + 1.49) = 2.58_5$
$pH = 11.41_5$

Mid1: $pH \approx pK_{a2} = 10.32$

End1: $pH \approx \frac{1}{2}(pK_{a2} + pK_{a1})$
$= \frac{1}{2}(10.32 + 6.37) = 8.34$

Mid2: $pH \approx pK_{a1} = 6.37$

End2: $pH \approx \frac{1}{2}(pc' + pK_{a1})$
$= \frac{1}{2}(1.61 + 6.37) = 3.99$

The pH titration curve is at the right.

d. Phenolphthalein is also well suited to the first endpoint of this tiration; however, the endpoint here is signified by pink → clear instead of the reverse.

57. The small required $[Cu^{2+}]$ combined with the large K_f for complexation ensure that the conversion of $[Cu^{2+}]$ to $[Cu(NH_3)_4^{2+}]$ is complete, and that we may use the equilibrium condition for the overall complexation reaction

$$Cu^{2+}(aq) + 4NH_3(aq) \rightleftharpoons [Cu(NH_3)_4^{2+}]$$

to find the required $[NH_3]$ as

$$[NH_3]^4 > \frac{[Cu(NH_3)_4^{2+}]}{K_f[Cu^{2+}]} = \frac{0.025}{(1 \times 10^{12})(1 \times 10^{-15})} = 25$$

or $[NH_3] > 2.24$ M.

59. See text key. When writing reaction quotients, omit pure condensed phases and solvents.

61. From Appendix C, $\Delta G_{298}^{\circ} = -0.82$ kcal/mol, and from Equation 12.8, $K = 4.0$. Extent-of-reaction analysis gives

	2HI(g) $\rightleftharpoons$	H$_2$(g) +	I$_2$(s)	Total pressure
				P
initial	P_0	0	...	P_0
equilibrium	$P_0 - 2x$	x	...	$P_0 - x$

Note that $I_2(s)$, a pure condensed phase, does not appear in Q, so it need not be included in the above analysis. The equilibrium condition is thus

$$K = \frac{P_{H_2}}{P_{HI}^2} = \frac{x}{(P_0 - 2x)^2}$$
$$x^2 - [P_0 + 1/(4K)]x + P_0^2/4 = 0$$
$$x^2 - 5.0626x + 6.2500 = 0$$

where in the last line we have substituted $P_0 = 5.00$ atm and $K = 4.0$. The chemically apropos root is $x = 2.134$ atm, yielding

$$P_{H_2} = 2.13 \text{ atm} \quad P_{HI} = 0.73 \text{ atm} \quad P = 2.86 \text{ atm}$$

As for the small vapor pressure exerted by $I_2(s)$ (see Figure 5.3), 0.5 torr at 298 K, not to worry. As long as some $I_2(s)$ is present, the solid-gas equilibrium fixes P_{I_2} at this constant value (see Chapter 14), and the above equilibrium calculation is unaffected. If P_0 were less than twice the vapor pressure of $I_2(s)$, however, no solid could form. The above equilibrium would then involve *only* $I_2(g)$, which would have to be included in the reaction quotient.

63. Appendix C gives $\Delta G_{298}^{\circ} = -0.123$ kcal/mol; Equation 12.8 then gives $K = 1.23$. The equilibrium condition then yields

$$[SO_2] = KP_{SO_2} = (1.23 \text{ M/atm})([0.050/760] \text{ atm}) = 8.1 \times 10^{-5} \text{ M}$$

For this small c and the none-too-small K_{a1} ($pK_{a1} = 1.81$ from Table 12.3), ionization should be nearly complete, yielding $[H^+] \approx c = 8.1 \times 10^{-5}$ M and pH = 4.09. Exact solution of the quadratic yields the same. Because of this situation, the equilibrium should actually be written for the ionized products $H^+(aq) + HSO_3^-(aq)$, with $K_{overall} = KK_{a1}$. For higher P_{SO_2} this becomes unnecessary. Note that P_{SO_2} is taken to be constant here; in a closed container depletion of the gas would have to be accounted for.

65. The reaction is $2NaHCO_3(s) \rightleftharpoons Na_2CO_3(s) + CO_2(g) + H_2O(g)$. The equilibrium condition is shown in Equation 12.54, where K is given as 4.7×10^{-7}. Equation 12.9 implies that the reaction is not spontaneous when $Q > K$. For the stated partial pressures we have

$$Q = P_{CO_2}P_{H_2O} = \frac{(0.38\,\text{torr})(3.0\,\text{torr})}{(760\,\text{torr}/\text{atm})^2} = 2.0 \times 10^{-6} > K$$

Therefore, provided a little decomposition has already occurred, and solid Na_2CO_3 is present, the baking soda is stable; in fact, it will reform under these conditions. Yes, breathing on it should help, since this builds up P_{CO_2} and increases Q. Low humidity, on the other hand, will reduce Q; the minimum required P_{H_2O} to maintain stability is determined by $Q = K$, or

$$P_{H_2O} = \frac{K}{P_{CO_2}} = \frac{4.7 \times 10^{-7}\,\text{atm}^2}{[0.38/760]\,\text{atm}} = 9.4 \times 10^{-4}\,\text{atm} \ (0.71\,\text{torr})$$

67. "Saturated" means that equilibrium has been reached, and we can therefore use our usual analysis. We have

$$Ca(OH)_2(s) \rightleftharpoons \underset{x}{Ca^{2+}(aq)} + \underset{2x}{2OH^-(aq)}$$

and

$$K_{sp} = [Ca^{2+}][OH^-]^2 = x(2x)^2$$

$$x = \sqrt[3]{K_{sp}/4} = \sqrt[3]{4.7 \times 10^{-6}/4} = 0.0105_5\,\text{M}$$

Thus $[Ca^{2+}] = 0.0105_5$ M, $[OH^-] = 0.0211$ M, and pH $= 14 - 1.68 = 12.32$.

69. The potential precipitation equilibrium, written conventionally in the reverse direction, is
$$Cu(OH)_2(s) \rightleftharpoons Cu^{2+}(aq) + 2OH^-(aq), \quad K_{sp} = [Cu^{2+}][OH^-]^2 = 5.5 \times 10^{-20}$$
from Table 12.6. Precipitation requires the ion product $Q_{ion} = [Cu^{2+}]_0[OH^-]_0^2 > K_{sp}$. Accounting for dilution of each solution upon mixing, we have
$$Q_{ion} = (0.0250)(0.100)^2 = 2.5 \times 10^{-4}$$
which exceeds K_{sp} by more than 15 orders of magnitude. Since NaOH is in stoichiometric excess, we carry out an extent of reaction analysis treating excess OH^- as a common ion, as discussed for the $PbCl_2$ example in the text below Equation 12.56. OH^- remaining is

$$\text{mmol OH}^- = (0.200\,\text{M})(50.0\,\text{mL}) - \left(\frac{2\,\text{mol OH}^-}{1\,\text{mol Cu}^{2+}}\right)(0.0500\,\text{M})(50.0\,\text{mL}) = 5.0\,\text{mmol}$$

and $[OH^-] = c = 5.0$ mmol / 100.0 mL $= 0.050$ M. The common ion analysis is

$$Cu(OH)_2(s) \rightleftharpoons Cu^{2+}(aq) + 2OH^-(aq)$$

$$x \qquad c + 2x \qquad K_{sp} = x(c+2x)^2 \approx xc^2$$

yielding $x = [Cu^{2+}] = (5.5\times10^{-20})/(0.050)^2 = 2.2\times10^{-17}$ M. The answer shows both that our neglect of the OH^- contribution from $Cu(OH)_2$ is justified, and that the precipitation is stoichiometric.

71. We write the overall equilibrium as

$$BaCO_3(s) + 2H^+(aq) \rightleftharpoons Ba^{2+}(aq) + H_2CO_3(aq)$$

This reaction, a heterogeneous Brønsted-Lowry acdi-base reaction, is the difference of the solubility equation of $BaCO_3(s)$, $K_{sp} = 2.6\times10^{-9}$ from Table 12.6, and the sum of the two acid ionization steps of H_2CO_3, as in Equation 12.49, $K_{12} = K_{a1}K_{a2} = 2.1\times10^{-17}$. Thus

$$K = K_{sp}/K_{12} = 1.26\times10^8$$

and the reaction is highly spontaneous. The stoichiometry demands $[Ba^{2+}] = [H_2CO_3]$, and thus the equilibrium condition determines the necessary pH:

$$K = \frac{[Ba^{2+}][H_2CO_3]}{[H^+]^2}$$

$$[H^+]^2 = \frac{[Ba^{2+}]^2}{K} = \frac{(0.10)^2}{1.26\times10^8} = 7.9\times10^{-11}$$

Therefore $[H^+] = 8.9\times10^{-6}$ M, and pH = 5.05. Below this pH precipitation is prevented, or the reaction as written is spontaneous. This method of solution combines into one step the two-step method suggested in the hint; from K_{sp} the free $[CO_3^{2-}]$ is 2.6×10^{-8} M, and $[H^+]$ then follows from Equation 12.49. The equilibrium $[H_2CO_3]$ exceeds the solubility of $CO_2(g)$ in water at 1 atm and 298 K, and thus the reaction will "fizz."

73. This exercise is similar to Exercise 71. The critical pH above which precipitation becomes spontaneous in saturated H_2S solution will differ from one sulfide to another according to the magnitude of K_{sp} for each. Qualitatively, the smaller K_{sp} is (that is, the less soluble the salt), the lower the pH must be to prevent precipitation. This coupling arises because low pH inhibits the ionization of H_2S, thereby reducing $[S^{2-}]$ and with it the likelihood of sulfide precipitation. From Table 12.6, $K_{sp}(NiS) = 9.2\times10^{-23}$ and $K_{sp}(CoS) = 8.2\times10^{-22}$, that is, NiS is less soluble than CoS, and therefore the pH needed to prevent NiS precipitation is lower than that for CoS for a given cation concentration. Quantitatively, we employ the combined two-step equilibrium condition of Equation 12.49, with the solubility equilibrium used to determine the equilibrium $[S^{2-}]$:

$$K_{a1}K_{a2} = \frac{[H^+]^2[S^{2-}]}{[H_2S]}$$

$$[H^+]^2 = \frac{K_{a1}K_{a2}[H_2S]}{[S^{2-}]} = \frac{K_{a1}K_{a2}[H_2S][M^{2+}]}{K_{sp}}$$

$$= \frac{(9.1\times10^{-8})(1.29\times10^{-14})(0.10)(0.050)}{K_{sp}}$$

$$= \frac{5.87\times10^{-24}}{K_{sp}}$$

where Table 12.3 data and the known concentrations are substituted in the third line. Substituting the above K_{sp}s in turn, we find

NiS: $[H^+] = 0.25$ M, pH = 0.60

CoS: $[H^+] = 0.085$ M, pH = 1.07

Thus, for pH between 0.60 and 1.07, NiS will remain insoluble while CoS dissolves, and the two ions can be successfully separated by this differential precipitation method.

75. We use Equation 12.60 with the van der Waals parameters for H_2 from Table 9.3, $a = 0.2421$ atm L^2/mol^2 and $b = 0.02651$ L/mol. In L atm units,

$$RT = (0.082057 \text{ L atm/K mol})(298.15 \text{ K}) = 24.465 \text{ L atm/mol at NTP}$$

Thus

$$\ln\gamma = -\frac{P}{RT}\left(\frac{a}{RT} - b\right)$$

$$= -\frac{P}{24.465 \text{ L atm/mol}}\left(\frac{0.2421 \text{ atm L}^2/mol^2}{24.465 \text{ L atm/mol}} - 0.02651 \text{ L/mol}\right)$$

$$= \frac{P}{1473 \text{ atm}}$$

For the choices of P given in the problem, we construct a small table:

P(atm)	$\ln\gamma$	γ	γP(atm)
1.00	6.79×10^{-3}	1.0007	1.00_{07}
10.0	6.79×10^{-2}	1.007	10.0_7
1000.	0.679	1.972	1972

We note that in the calculation above, $b > a/RT$, indicating the dominance of repulsive force. This makes $\gamma > 1$, endowing compressed H_2 with more free energy than that due to its pressure alone; compressed H_2 is capable of more PV work than an ideal gas, since the H_2 molecules are pushing each other apart, and this repulsive energy is released upon expansion. Both Equation 12.60 and the calculations show that this contribution to G increases linearly with increasing P.

77. As in Exercise 75, we use Equation 12.60, now with the van der Waals constants for H_2O from Table 9.3, $a = 5.463$ atm L^2/mol^2 and $b = 0.03048$ L/mol. At the two temperatures $T_1 = 373.15$ K and $T_2 = 473.15$ K, we have

$$RT_1 = (0.082057 \text{ L atm/K mol})(373.15 \text{ K}) = 30.620 \text{ L atm/mol}$$

$$RT_2 = (0.082057 \text{ L atm/K mol})(473.15 \text{ K}) = 38.825 \text{ L atm/mol}$$

Equation 12.60 then yields

$$\ln\gamma = -\frac{P}{RT}\left(\frac{a}{RT} - b\right)$$

$$= -\frac{1.00\text{ atm}}{RT}\left(\frac{5.463\text{ atm L}^2/\text{mol}^2}{RT} - 0.03048\text{ L}/\text{mol}\right)$$

$$\ln\gamma_1 = -0.03266(0.17842 - 0.03048) = -0.004831$$

$$\ln\gamma_2 = -0.02576(0.14071 - 0.03048) = -0.002839$$

and $\gamma_1 = 0.995$, $\gamma_2 = 0.997$. Thus steam is nearly ideal under these conditions, and becomes more so with increasing T. As shown in the calculations, at these Ts $a/RT > b$, indicating dominance of attractive forces between water molecules, and making $\gamma < 1$. At still higher T, repulsive forces will take over, as they already have in H_2 at NTP (see Exercise 75). At the *Boyle temperature* $T_B = a/bR$ the attractive and repulsive corrections cancel, and the free energy of the gas (as well as its other gaseous properties) reverts to (nearly) ideal behavior.

13

Electrochemistry

13 Electrochemistry

<u>Your Chapter 13 GOALS</u>:

- Identify electrical work as a second form of work in addition to *P-V* work
- Observe that many redox reactions have large $\Delta G°$ and K values
- Define a volt and potential difference
- Use the Nernst equation to determine cell potentials
- Relate cell voltage to the free energy change of a reaction
- Distinguish between electrolytic and voltaic (or galvanic) cells
- Diagram electrolytic and voltaic cells
- Detail the workings of the four voltaic cells in Section 13.2
- Utilize half-reactions to obtain overall cell reactions, as well as $\mathscr{E}°$ and $\Delta G°$ values
- Relate standard reduction potentials to standard electrodes such as the standard hydrogen electrode
- Relate cell voltage to pH for the pH electrode
- Use Faraday's Laws of electrolysis to calculate charge passed, current, and times for electrolytic processes
- Outline the key aspects of electrolytic cells
- Use half-reactions to detail the process of corrosion

<u>Chapter 13 KEY EQUATIONS</u>:

- 13.3, 13.4, 13.6, 13.7, 13.15, 13.16, 13.19

<u>Overview</u>

Chapter 13 has a clear connection to the prior chapters in *University Chemistry*. K, pH, $\Delta G°$, and work are topics of concern. Table 13.1 shows that metals easily lose electrons, while many non-metals (F_2, Cl_2, Br_2) easily gain electrons. Many other examples exists, but there is a uniqueness to Chapter 13, all of which centers around the volt. Note that this unit is not used in other areas of the book. To understand electrochemistry, however, you must grasp the concept of voltage, or (electric) potential difference. As the text says, the electrons are "pushed" through an external circuit by the electric potential difference. Equation 13.4 indicates that cell reactions with a large ΔG will give the electrons in the external circuit a stronger push than those cells with a small ΔG. This is a reflection of the degree of spontaneity that we have seen previously for negative ΔG values. Section 13.1 contains the Nernst Equation, Equation 13.5, and you should note that its form is very similar to Equation 12.6, due to Equation 13.4. See Exercise 4. For Exercise 2 note that power and the watt were defined in Exercise 4 of Chapter 2 and that Equation 13.19 is needed. To solve Exercise 6, you must calculate K for the reverse of Equation 5.8 and use $K = [Cu^{2+}]/[Zn^{2+}]$. This Exercise also reminds you that, when a reaction is reversed, the sign of $\Delta G°$ and $\mathscr{E}°$ must be switched.

Section 13.2 is a survey of the practical applications of redox electrochemistry. The cells in the Section work *spontaneously*; no outside voltage source is needed to make the electrons flow. This is the definition of a voltaic cell, or if you don't like frogs, a galvanic cell. Any cell, voltaic or electrolytic, must have an electron source and sink. As noted in the text, in Figure 13.2 the Cu anode is the source of electrons *within the cell*. From the viewpoint of a battery's ability to provide electrons, that is, *outside of the cell*, it is that the Zn anode that is the source of electrons. For this reason the (−) sign on a battery indicates that electrons leave the battery from that end. See Exercise 7. The Zn *anode* within the cell, however, must be positive because *anions* flow toward it. Conversely, the Cu *cathode* attracts *cations*, and it must be negative within the cell.

Compare these signs carefully with the electrolytic cells of Figures 13.11 and 13.12, where the anode is now given a (+) sign outside of the cell. The (−) sign is placed on the cathode, the point at which the outside voltage source provides electrons to the cell.

For Exercise 10 remember that two moles of electrons are required to reduce 1 mol of Cu^{2+} and use Equation 13.19. The cell of Exercise 12 is expected to provide a more constant voltage for two reasons. First, the alkaline paste conducts ions more effectively than the paste in the Leclanché cell. Second, the overall cell reaction for the alkaline dry cell contains only solids, meaning that Q of Equation 13.6 is one and that $\mathscr{E} = \mathscr{E}°$ throughout the life of the battery. Exercise 18 is made rather difficult because use of the Nernst equation is not straightforward due to the complexity of the sulfuric acid equilibrium that is shown in Example 13.3, where we have shown the sulfuric acid completely dissociated. Some chemists write the reaction in this manner, yet others write the H_2SO_4 dissociation as $2H^+ + 2HSO_4^-$. This approach acknowledges K_{a2} for sulfuric acid is not nearly as large as K_{a1}, which was covered in Exercise 47 of Chapter 12. The correct way to treat the situation, however, is to use the equilibrium considerations of Chapter 12. We circumvent that difficult problem by simply using $Q = 1/[H_2SO_4]^2$, which yields the answers in the text answer key. When solving Exercises 20 and 21, assume that the *molten* cases discussed therein are entropicly similar to the aqueous case, allowing you to use the *aqueous* standard reduction potentials of Table 13.1. Exercises 20 and 21 also remind you that Equation 13.15 reflects the intensive nature of cell potentials, meaning that you do *not* multiply the voltages that you obtain from Table 13.1 by the stoichiometric coefficients for use in Equation 13.15 no matter what must be done to get the half-reactions to balance.

Hopefully, you are now familiar with the process of selecting standard reference states as the "zero" of a particular property, as we have done for $\Delta H°_f$ and $\Delta G°_f$ and, *for solutions*, $S°$. In Section 13.3 the standard hydrogen electrode is set as the zero of standard reduction potentials. Therefore, the voltage ladder of Figure 13.7 is quite analogous to the enthalpy level diagrams we constructed in Chapter 10; it is only the difference between the levels that we may know. For Exercise 26, then, note that the Zn^{2+}/Zn couple is 0.76 V *below* the SHE on the voltage ladder. Therefore, the unknown couple must be 0.77 V *above* the SHE. Exercises 36, 37, and 38 also look at the voltage ladder concept, which relates standard reduction potentials. The overall $\mathscr{E}°$'s for Exercises 36 and 37 can be added to obtain $\mathscr{E}°$ for the cell of Exercise 38. Exercise 22 affords us the chance to contrast the intensive cell potential seen in Exercises 20 and 21, among many other Exercises in the chapter, with the *extensive* free energy change for the overall cell reaction. For the aluminum-chlorine cell, one must multiply $\Delta G°_{298}$ for the anode half-reaction by 2 and the cathode half-reaction by 3. The overall free energy change for the reaction is, therefore, found by summing 232 kcal and − 188.232 kcal. This crucial difference between $\mathscr{E}$ and ΔG is the reason that an *n* appears in Equation 13.4. Exercises 27 and 28 are very similar to the $Al(OH)_3$ example at the end of Section 13.3, and they can be solved in the same manner – by using Equation 13.4 and Table 13.1 for the metal reduction and Equation 12.8 for the K_{sp} expression. Exercises 29 and 32 are very similar to this, only K_w is needed instead of K_{sp}. We want to reiterate, however, that you can also solve these Exercises by first determining the $\Delta G°$ values for both half-reactions from Appendix C and Equation 12.2, followed by adding the half-reaction $\Delta G°$'s to obtain the overall $\Delta G°$. (The intimate relationship between $\mathscr{E}$ and ΔG is responsible for this.) The steam reforming reaction of Exercise 34 is obtained by adding the overall reaction of the CH_4/O_2 cell, $CH_4(g) + 2O_2(g) \rightarrow CO_2(g) + 4H_2(g)$, with the reverse of the overall reaction for the H_2/O_2 cell, $2H_2O(l) \rightarrow 2H_2(g) + O_2(g)$. Note that the text answer key contains the results for $H_2O(l)$ and $H_2O(g)$.

Exercise 40 shares concepts with the entropy of mixing idea that we examined in Exercise 16 of Chapter 11 and the ΔG expression that was derived in Exercise 28 of Chapter 11. The key,

13 Electrochemistry

then, is to assume that $\Delta H = 0$ because the ideal solutions will merely mix and not react or exert forces on each other, just as is assumed with gases in kinetic molecular theory. Therefore, solvent/ion "bonds" are ignored. After this assumption use Equations 11.24 and 13.4.

Electrolysis makes the impossible possible. An applied voltage "pushes" electrons in a direction opposite to where they would go if they were simply responding to the free energy "push" that drives them in the spontaneous direction. This process is responsible for obtaining active metals which are found naturally as ions, having already undergone their typical oxidation. For Exercise 44 note that you will need to account for the fact that $\mathscr{E}° = -2.06$ V, whereas $\mathscr{E} = -1.229$ V. We again urge you to consider what the ° symbol means. The -1.229 V value arises from a solution that initially has almost no H^+ or OH^-. When starting Exercise 48, determine how many moles of electrons the appropriate silver electrode loses. You will need Equation 13.19 for Exercises 47 through 50.

1. The stoichiometry of Equation 13.1 gives

$$g(Zn) = 0.500 \, L\left(\frac{0.100 \, \text{mol Ag}^+}{L}\right)\left(\frac{1 \, \text{mol Zn}}{2 \, \text{mol Ag}^+}\right)\left(\frac{65.39 \, g}{1 \, \text{mol}}\right) = 1.63 \, g \, Zn$$

For any finite starting concentrations, the large K value indicates that the reaction will truly go to completion. So, from the 2:1 stoichiometry 0.0500 mol Ag^+ will form 0.0250 mol Zn^{2+}. With no volume change this gives $[Zn^{2+}] = 0.0250 \, \text{mol}/0.500 \, L = 0.0500 \, M$. Equation 13.2 then gives

$$[Ag^+] = \sqrt{\frac{0.0500}{6.1 \times 10^{52}}} = \frac{9.054 \times 10^{-28} \, \text{mol}}{L}\left(\frac{6.0221 \times 10^{23} \, \text{ions}}{\text{mol}}\right) = \frac{5.5 \times 10^{-4} \, \text{ions}}{L}$$

Lastly, the system is 0.500 L. Therefore, there are 2.7×10^{-4} ions present. As in the text example with this problem, this is a nonsense answer, but we still predict that there will be no Ag^+ ions in solution at equilibrium.

3. a. The Faraday links the number of electrons and charge. With $Q = (0.833 \, \text{C/s})(3600 \, s) = 3.00 \times 10^3$ C from Exercise 2,

$$N_e = 3.00 \times 10^3 \, C\left(\frac{1 \, \text{mol e}^-}{96,485 \, C}\right)\left(\frac{6.0221 \times 10^{23}}{1 \, \text{mol}}\right) = 1.87 \times 10^{22} \, e^-$$

Dividing by N_A gives 0.031 mol e^-.

  b. In each cell we need one mole of electrons to form one mole of Ag. Thus,

$$g(Ag) = 3.00 \times 10^3 \, C\left(\frac{1 \, \text{mol e}^-}{96,485 \, C}\right)\left(\frac{1 \, \text{mol Ag}}{1 \, \text{mol e}^-}\right)\left(\frac{107.868 \, g}{1 \, \text{mol}}\right) = 3.35 \, g \, Ag$$

5. At equilibrium $\Delta G = 0$ and $\mathscr{E} = 0$ via Equation 13.4. Thus, Equation 13.6 reduces to Equation 13.7 since $K = Q_{eq}$. From Appendix C and Equations 13.4 and 13.7

$$K = \log^{-1}\left(2\left(\frac{72.01 \, \text{kcal}}{2 \, \text{mol}(23.0605 \, \text{kcal/mol V})}\right)\bigg/0.05916 \, V\right) = 6.1 \times 10^{52}$$

7. As in Figures 13.2, 13.5, and 13.6, the anode is labeled as (–), and the cathode is labeled (+). Thus, the Zn electrode is (–), and the Ag electrode is (+). The sketch is like Figure 13.2, except the cathode reaction is $Ag^+ + e^- \rightarrow Ag$, $AgNO_3$ is the cathode electrolyte, the salt bridge ions are Na^+ and Cl^-, and the voltmeter would read 1.56 V from $Zn + 2Ag^+ \rightarrow Zn^{2+} + 2Ag$.

9. The ionic equation for Volta's cell is Equation 13.1. Now, Appendix C gives $\Delta H^\circ_{298} = -87.25$ kcal and $\Delta S^\circ_{298} = -0.0512$ kcal/K. At 323 K, Equation 11.24 gives $\Delta G^\circ_{323} = -70.7$ kcal. Equation 13.4 then gives $\mathscr{E}^\circ = 70.7 \, \text{kcal}/[2 \, \text{mol}(23.0605 \, \text{kcal/mol V})] = 1.533 \, V$. We finally substitute into Equation 13.5 to get

$$\mathscr{E} = 1.533 \, V - \frac{1.9872 \, \text{cal/K mol}(323 \, K)}{2(23060.5 \, \text{cal/mol V})} \ln\frac{[Zn^{2+}]}{[Ag^+]^2} = 1.533 \, V - 0.0139\ln\frac{[Zn^{2+}]}{[Ag^+]^2}$$

Finally, using $\ln x = 2.3026\log x$,

$$\mathscr{E} = 1.533 \, V - 0.0320\log\frac{[Zn^{2+}]}{[Ag^+]^2}$$

In the above derivation, the temperature (323 K) was needed to calculate both terms. Thus, both terms are temperature dependent.

11. The electrons are already balanced. Thus, $Zn(s) + 2MnO_2(s) + 2NH_4^+(aq) \rightarrow Zn^{2+}(aq) + Mn_2O_3(s) + 2NH_3(aq) + H_2O(l)$. Note that it is the Mn that is reduced at the cathode. The oxidation at the anode is easy to see. To use Equation 13.15, we use, by convention, the standard *reduction* potential from Table 13.1 for the Zn half-cell even though the Zn is oxidized. Thus, $1.50\ V = \mathscr{E}°_{cath} - (-0.76\ V)$ yields $\mathscr{E}°_{cath} = 0.74\ V$. When writing the line notation, note that the cathode electrolyte solution contains three compounds and that the actual electrode is an inert carbon rod. So, we list the reduction product as the cathode along with the afore mentioned carbon. The line notation is

$$Zn\,|\,Zn^{2+}\,||\,NH_4^+,MnO_2,NH_3\,|\,Mn_2O_3(C)$$

13. The overall cell reaction is $Zn(s) + 2MnO_2(s) \rightarrow ZnO(s) + Mn_2O_3(s)$. Now, if the battery is completely new, there are no products. That is, all of the 70.0 % is due to the two reactants, Zn and MnO_2. If we let x = the mols of each reactant,

$$x\left(\frac{65.39\ g}{mol}\right) + 2x\left(\frac{86.9368\ g}{mol}\right) = 23.0\ g\,(0.700)$$

using $2x$ for MnO_2 because of the reaction stoichiometry. This gives $x = 0.06729$ mol. Zn is the anode, and the amount of charge that the cell can deliver is determined by its mass:

$$charge = 0.06729\ mol\ Zn\left(\frac{2\ mol\ e^-}{1\ mol\ Zn}\right)\left(\frac{96,485\ C}{mol\ e^-}\right) = 1.30 \times 10^4\,C$$

Strictly speaking, the charge drawn is negative because the charge is due to electrons. (Often, this is simply ignored or understood. See Exercise 2.) Thus, assuming that 1.5 V is an exact number, $w_{max} = Q\mathscr{E} = (-1.30 \times 10^4\ C)(1.5\ J/C) = -19,500\ J = -19.5\ kJ = -4.66$ kcal. (See also Exercise 10.) The work is to be done by the system, so it is negative. From Equation 13.19, $t = Q/I = 1.30 \times 10^4\ C/(0.050\ C/s) = 2.6 \times 10^5\ s \approx 72$ hours $\approx$ three days.

15. By comparing Equations 13.11 and 13.13, we see that in basic medium water is produced at the anode and hydroxide ion is produced at the cathode. Analogously,

$$Zn(s) + 2OH^-(aq) \rightarrow ZnO(s) + H_2O(l) + 2e^- \qquad \text{Anode } (-)$$
$$HgO(s) + H_2O(l) + 2e^- \rightarrow Hg(l) + 2OH^-(aq) \qquad \text{Cathode } (+)$$

$$\overline{Zn(s) + HgO(s) \rightarrow ZnO(s) + Hg(l)}$$

As usual the anode is $(-)$. For the anode Appendix C gives $\Delta G°_{298,a} = -57.58$ kcal $(\approx -240.0\ kJ)$, and Equation 13.4 gives $\mathscr{E}°_{anod} = 57.58$ kcal/[2 mol(23.0605 kcal/mol V)] = 1.248 V. For the cathode Appendix C gives $\Delta G°_{298,c} = -4.506$ kcal $(\approx -18.85\ kJ)$, and Equation 13.4 gives $\mathscr{E}°_{cath} = 4.506$ kcal/[2 mol(23.0605 kcal/mol V)]= 0.0977 V. So, $\Delta G°_{298,c} + \Delta G°_{298,c} = -62.09$ kcal. Equation 13.4 then gives $\mathscr{E}°_{cell} = 62.09$ kcal/[2 mol(23.0605 kcal/mol V)] = 1.346 V. To use Equation 13.15, we have to change the sign of $\mathscr{E}°_{anod}$ because we have written the reaction and calculated it as an oxidation. By convention, though, we use the *reduction* potential in Equation 13.15. Of course, calculating $\Delta G°_{298}$ for the overall reaction will yield the same result. See text key for line notation. The figure is at the right.

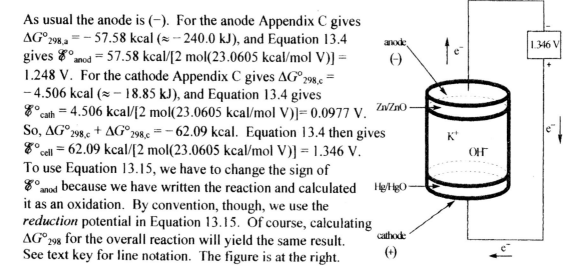

17. It is important to note a key difference in this cell from the Leclanché dry cell, the alkaline dry cell, and the "button batteries" of Exercises 15 and 16. In all of those cells Zn was oxidized (at the anode). Table 13.1 reveals, however, that the oxidation of Li will clearly be favored over the oxidation of Zn. Conversely, it is much more difficult to reduce Li than to reduce Zn. See the text key for the line notation and notice that Zn is now on the right side. What does remain the same, however, is that in a basic medium H_2O forms at the anode and OH^- at the cathode, along with the other given products. Thus,

$$2Li(s) + 2OH^-(aq) \rightarrow Li_2O(s) + H_2O(l) + 2e^- \qquad \text{Anode } (-)$$
$$ZnO(s) + H_2O(l) + 2e^- \rightarrow Zn(s) + 2OH^-(aq) \qquad \text{Cathode } (+)$$

$$2Li(s) + ZnO(s) \rightarrow Li_2O(s) + Zn(s)$$

As usual, the anode is (−). For the anode Appendix C gives $\Delta G°_{298,a} = -115.63$ kcal (≈ -483.80 kJ), and Equation 13.4 gives $\mathscr{E}°_{anod} = 115.63$ kcal/[2 mol(23.0605 kcal/mol V)] = 2.507 V. For the cathode Appendix C gives $\Delta G°_{298,c} = 57.58$ kcal. From Equation 13.4 the standard half-cell potential at the cathode is $\mathscr{E}°_{cath} = -57.58$ kcal/[2 mol(23.0605 kcal/mol V)] = −1.248 V. So, $\Delta G°_{298,a} + \Delta G°_{298,c} = -58.05$ kcal. Equation 13.4 then gives 58.05 kcal/[2 mol(23.0605 kcal/mol V)] = 1.259 V = $\mathscr{E}°_{cell}$. If you want to use Equation 13.15, see the note in Exercise 15's solution.

19. Example 13.3 gives the lead-acid battery overall cell reaction for which Appendix C yields $\Delta G°_{298} = -94.20$ kcal and Equation 13.4 gives $\mathscr{E}°_{298} = 2.042$ V. To calculate $\Delta G°_{258}$, we use Appendix C to get $\Delta H°_{298} = -75.43$ kcal and $\Delta S°_{298} = 0.0630$ kcal/K, yielding $\Delta G°_{258} = -91.7$ kcal via Equation 11.24. Equation 13.4 then gives $\mathscr{E}°_{258} = 91.7$ kcal/[2 mol(23.0605 kcal/mol V)] = 1.988 V. For one cell, then, the voltage drop is the difference between the potentials at the two different temperatures. That is, 0.055 V. For six cells, the voltage drop is 6(0.055 V) = 0.33 V. This is a large enough drop, perhaps, to hinder starting the car.

21. Sodium is more easily oxidized (harder to reduce) than sulfur. Thus, we expect the sodium to be oxidized and the sulfur to be reduced. This leads to the following half-reactions:

$$[Na \rightarrow Na^+ + e^-] \times 2 \qquad \text{Anode } (-)$$
$$S + 2e^- \rightarrow S^{2-} \qquad \text{Cathode } (+)$$

$$2Na + S \rightarrow 2Na^+ + S^{2-}$$

Assuming that the molten metal case can be treated as if the reaction occurred in aqueous solution and using Equation 13.15 gives $\mathscr{E}°_{cell} = -0.51$ V $- (-2.71$ V) = 2.20 V. Remember that even though we need to multiply the Na half-reaction by 2 to get the overall cell reaction, we do not multiply the $\mathscr{E}°$ value from Table 13.1. The overall cell reaction for the solid state case is $2Na(s) + S(s) \rightarrow Na_2S(s)$, and Appendix C immediately gives $\Delta G°_{298} = -83.6$ kcal. Equation 13.4 gives $\mathscr{E}°_{solid} = 83.6$ kcal/[2 mol(23.0605 kcal/mol V)] = 1.81 V. (Of course, the cell does not operate at 298 K since the reactants must be molten.)

23. See text key for half-reactions. If $\mathscr{E}°_{cell} > 0$, the reaction is spontaneous as written. If $\mathscr{E}°_{cell} < 0$, the reverse reaction is spontaneous. Remember that even if we have to multiply one or both of the reactions by a factor to get the electrons to cancel, we do NOT change the value of $\mathscr{E}°$ for the half reaction. The calculations are:

    a. $\mathscr{E}°_{cell} = -0.13$ V $- (-0.23$ V) = 0.10 V

    b. $\mathscr{E}°_{cell} = 0.80$ V $- (-0.41$ V) = 1.21 V

c. $\mathscr{E}^\circ_{cell} = \quad 0\ V \ - \ (-2.36\ V) \ = \ 2.36\ V$

d. $\mathscr{E}^\circ_{cell} = -0.14\ V \ - \ (1.087\ V) \ = \ -1.23\ V$

e. $\mathscr{E}^\circ_{cell} = \ 1.360\ V \ - \ (0.535\ V) \ = \ 0.825\ V$

Thus, all except (d) are spontaneous as written. (d) is spontaneous in the reverse direction.

25. The easiest way to solve this problem is to remember that, when element or compound A in Table 13.1 has a higher (more +) standard reduction potential than element or compound B, A's reduction will occur before B's. So, comparing Ag(s) and $H^+(aq)$, shows that the reduction of Ag(s) will occur preferentially to that of $H^+(aq)$. In other words, $H^+(aq)$ is not a strong enough oxidizing agent to oxidize Ag(s). Therefore, Ag(s) will not react with $H^+(aq)$. The same goes for any other metal above $H^+(aq)$. For this problem that means Ag(s) and Cu(s) will not react with $H^+(aq)$. By the same reasoning the metals that lie below $H^+(aq)$ will react with acid. So, Fe(s), Mn(s), and Ni(s) will react with acid. Don't be confused by the Fe^{3+}/Fe^{2+} couple that is above $H^+(aq)$. This couple does not contain elemental iron. The Fe^{2+}/Fe couple does, and it is *below* 0 V.

27. The $Al(OH)_3$ example at the end of Section 13.3 is very similar to this problem. By analogy we add the following reactions:

$$Ni^{2+}(aq) + 2e^- \rightarrow Ni(s)$$
$$Ni(OH)_2(s) \rightarrow Ni^{2+}(aq) + 2OH^-(aq)$$

$$Ni(OH)_2(s) + 2e^- \rightarrow Ni(s) + 2OH^-(aq)$$

The Ni^{2+} reduction gives $\Delta G^\circ = -nF\mathscr{E}^\circ = -2\ mol(23.0605\ kcal/mol\ V)(-0.23\ V) = 10.6$ kcal. For the ionization of $Ni(OH)_2(s)$ Table 12.6 helps us to write $\Delta G^\circ = -nRT\ln K_{sp} = -1\ mol(0.0019872\ kcal/K\ mol)(298\ K)\ln(5.6 \times 10^{-16}) = 20.8$ kcal. (We have retained one more significant figure than is allowed.) Adding these free energies to give the free energy for the overall reaction gives $\Delta G^\circ_{tot} = 31.4$ kcal. Lastly, Equation 13.4 gives $\mathscr{E}^\circ = -31.4$ kcal/[2 mol(23.0605 kcal/mol V)] = -0.681 V. (We could also have calculated ΔG° for each part from Appendix C.)

29. The reaction sum is given in the problem as

$$2H^+(aq) + 2e^- \rightarrow H_2(g)$$
$$[H_2O(l) \rightarrow H^+(aq) + OH^-(aq)] \times 2$$

$$2H_2O(l) + 2e^- \rightarrow H_2(g) + 2OH^-(aq)$$

This implies that we need more free energy to reduce water than to reduce $H^+(aq)$ and that $\mathscr{E}^\circ$ will be negative since $\mathscr{E}^\circ = 0$, by convention, for the SHE reduction. Equation 13.4 gives $\Delta G^\circ = 0$ for the first reaction. The autoionization gives $\Delta G^\circ = -1\ mol(0.0019872$ kcal/K mol)(298 K)$\ln(1.0 \times 10^{-14}) = 19.1$ kcal. To get the free energy change for the overall reaction, we need to multiply the free energy change for the second half-reaction by two because free energy is an extensive property. Thus, $\Delta G^\circ_{tot} = 0 + 2(19.1$ kcal) = 38.2 kcal. Lastly, Equation 13.4 gives $\mathscr{E}^\circ = -38.2$ kcal/[2 mol(23.0605 kcal/mol V)] = -0.828 V. Thus, our tabulated value agrees with our prediction.

31. The key is to recall that the SHE potential has $\mathscr{E}^\circ = 0$, written in either direction. So, we can add the oxygen reduction half-reaction to the SHE and get the value of the oxygen reduction half-reaction back out. Their sum is the combustion of hydrogen reaction. Thus,

$$O_2(g) + 4H^+(aq) + 4e^- \rightarrow 2H_2O(l)$$
$$[H_2(g) \rightarrow 2H^+(aq) + 2e^-] \times 2$$

$$O_2(g) + 2H_2(g) \rightarrow 2H_2O(l)$$

By convention, only the first reaction contributes to the overall free energy, and $\Delta G^\circ_1 = \Delta G^\circ_{tot} = -4 \text{ mol}(23.0605 \text{ kcal/mol V})(1.229 \text{ V}) = -113.4 \text{ kcal}$.

33. For the cathode reaction of Equation 13.13

$$Q = \frac{[OH^-]^4}{P_{O_2}}$$

With $\mathscr{E}^\circ$ from Table 13.1, the Nernst Equation becomes

$$\mathscr{E} = 0.401 \text{ V} - \frac{0.05916}{4} \log \frac{[OH^-]^4}{P_{O_2}}$$

Of course, using ΔG°_f's from Appendix C with Equation 13.4 will give the same value. Simple substitution into this formula gives the text key values, but you must convert the Torr values into atmospheres by dividing by 760 because we have chosen the atmosphere as our standard unit of pressure way back in Equation 12.4.

35. We want the anode and cathode half-reactions to sum to $2CH_3OH(l) + 3O_2(g) \rightarrow 2CO_2(g) + 4H_2O(l)$, the balanced methanol combustion reaction. You can obtain the anode half-reaction by sight or by subtracting the cathode reaction from the overall reaction. The appropriate half-reaction sum is

$$[O_2(g) + 2H_2O(l) + 4e^- \rightarrow 4OH^-(aq)] \times 3 \qquad \text{Cathode}$$
$$[CH_3OH(l) + 6OH^-(aq) \rightarrow CO_2(g) + 5H_2O(l) + 6e^-] \times 2 \quad \text{Anode}$$

$$2CH_3OH(l) + 3O_2(g) \rightarrow 2CO_2(g) + 4H_2O(l)$$

From Appendix C $\Delta G^\circ_{cath} = 3(-37.002) = -111.006$ kcal and $\Delta G^\circ_{anod} = 2(-112.36_5) = -224.73$ kcal. (We could have used Table 13.1 and Equation 13.4.) Thus, $\Delta G^\circ_{cell} = \Delta G^\circ_{cath} + \Delta G^\circ_{anod} = -335.74$ kcal. So, from Equation 13.4 we get $\mathscr{E}^\circ_{cell} = 335.74$ kcal/[12 mol (23.0605 kcal/mol V)] = 1.213 V. To get the standard *reduction* potential for the anode, as in Table 13.1 (note that the above anode reaction is an oxidation), we reverse the reaction, which reverses the sign of ΔG°_{anod} to give $\Delta G^\circ_{anod} = 112.37$ kcal. Thus, Equation 13.4 gives $\mathscr{E}^\circ_{anod} = -112.37$ kcal/[6 mol(23.0605 kcal/mol V)] = -0.812 V. We could have calculated $\mathscr{E}^\circ_{cell}$ from Equation 13.15, too. That is, $0.401 - (-0.812 \text{ V}) = 1.213$ V.

37. When analyzing an electrochemical cell, the couple that lies higher in Table 13.1 will always become the cathode. Thus,

$$2H^+(aq) + 2e^- \rightarrow H_2(g) \qquad \text{Cathode}$$
$$Mg(s) \rightarrow 2Mg^{2+}(aq) + 2e^- \qquad \text{Anode}$$

$$Mg(s) + 2H^+(g) \rightarrow 2Mg^{2+}(aq) + H_2(g)$$

Equation 13.15 gives $\mathscr{E}^\circ_{cell} = 0 \text{ V} - (-2.36 \text{ V}) = 2.36$ V.

39. For the first question, the key word is standard. Standard concentration is 1 M. Thus, pH = $-\log(1) = 0$. The overall cell reaction is $H_2(g) + Cl_2(g) \rightarrow 2H^+(aq) + 2Cl^-(aq)$, which leads to the reaction quotient

$$Q = \frac{[H^+]^2[Cl^-]^2}{P_{H_2}P_{Cl_2}}$$

If, as the question says, all of the values are standard (1 M or 1 atm) except for H^+, we get $Q = [H^+]^2$. So, with $n = 2$ and Example 13.4 giving $\mathscr{E}° = 1.360$ V, the Nernst Equation is $\mathscr{E} = 1.360$ V $- 0.05916\log[H^+]$, using the fact that $\log x^a = a\log x$. Now, since pH $= -\log[H^+]$, $\mathscr{E} = 1.360$ V $+ 0.05916$pH. When pH $= 0$, then $\mathscr{E} = \mathscr{E}°$. The other three concentrations give pH's of 1, 7, and 14. Direct substitution into our derived equation leads to $\mathscr{E}$'s of 1.419 V, 1.774 V, and 2.188 V.

41. We are told to presume that the pH of the unknown is greater than zero. This means that the SHE will have the higher $[H^+]$, since its pH is zero. Thus, as in a concentration cell, there will be a lowering of $[H^+]$ in the SHE. This is analogous to the Cu/Cu^{2+} cell that is discussed in the text. The lowering of $[H^+]$ occurs by the reduction: $2H^+(aq) + 2e^- \rightarrow H_2(g)$, indicating that the SHE is the cathode. That is, the more concentrated cell becomes the cathode. The line notation is given in the text key. Note that the cell with the unknown concentration is, indeed, the anode.

As explained in Section 13.4, $\mathscr{E}° = 0$ for any concentration cell. To determine n and Q for the Nernst Equation we set up the following half-reactions:

$2H^+(aq) + 2e^- \rightarrow H_2(g)$ Cathode
$H_2(g) \rightarrow 2H^+(aq) + 2e^-$ Anode

$2H^+_c(aq) + H_{2,a}(g) \rightarrow H_{2,c}(g) + 2H^+_a(aq)$

This is analogous to the Cu concentration cell reaction in Section 13.4. We have included the subscripts to clearly express Q, the reaction quotient. Thus,

$$Q = \frac{P^c_{H_2}[H^+]^2_a}{P^a_{H_2}[H^+]^2_c}$$

If we assume that the pressures are 1 atm and recall that $[H^+]_c = 1$ M, this reduces to $Q = [H^+]^2_a$. Therefore, with $n = 2$, the Nernst Equation becomes $\mathscr{E} = -0.05916\log[H^+]_a$, using the fact that $\log x^a = a\log x$. Since pH $= -\log[H^+]$, the final Equation is $\mathscr{E} = 0.05916$pH$_a$. Direct substitution into this equation with pH$_a = 4.00$, 7.00, and 10.00 gives $\mathscr{E} = 0.237$ V, 0.414 V, and 0.592 V.

43. The reaction in the half-cell where the titration actually occurs is $Fe^{2+}(aq) \rightarrow Fe^{3+}(aq) + e^-$, indicating that $n = 1$ for the Nernst Equation. This is the reverse of the reduction reaction listed in Table 13.1. Thus, for the oxidation $\mathscr{E}° = -0.77$ V. The last thing we need for the Nernst Equation is Q, which is $[Fe^{3+}]/[Fe^{2+}]$ from the above half-reaction. The working form of the Nernst Equation for this Exercise, therefore, is

$$\mathscr{E} = -0.77 \text{ V} - 0.05916\log\left(\frac{[Fe^{3+}]}{[Fe^{2+}]}\right)$$

The reaction stoichiometry is 1:1. So, ¼ of the way through the titration we will have consumed ¼(0.100 M) $= 0.025$ M of Fe^{2+}, leaving 0.075 M $= [Fe^{2+}]$ and 0.0251 M $= [Fe^{3+}]$. At the half-way point the concentrations are essentially equal, and $\mathscr{E} \approx \mathscr{E}° = -0.77$ V, the standard potential of this oxidation reaction. This is what we saw in acid/base titrations and in buffers. Likewise, the following table is constructed:

$[Fe^{2+}]$	$[Fe^{3+}]$	$\mathscr{E}\,(V)$
0.100	0.00010	-0.59
0.075	0.02510	-0.74
0.050	0.0501	-0.77
0.025	0.0751	-0.80
0.0001	0.100	-0.95

The plot is below. The similarity to the buffer region in the titration curve of a weak base by an acid is apparent.

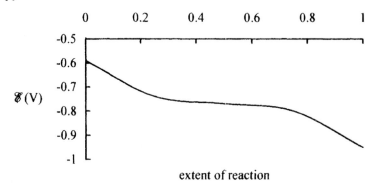

extent of reaction

45. $H_2(g)$ and $Cl_2(g)$ are produced. In the text example for the electrolysis of the NaI solution, it was shown that the cathode is the site of H_2 production. (See also Exercise 44.) Thus, the two half-reactions are

$$2H_2O(l) + 2e^- \rightarrow 2OH^-(aq) + H_2(g) \qquad \text{Cathode}$$
$$2Cl^-(aq) \rightarrow Cl_2(g) + 2e^- \qquad \text{Anode}$$
$$\overline{}$$

$$2H_2O(l) + 2Cl^-(aq) \rightarrow Cl_2(g) + 2OH^-(aq) + H_2(g)$$

We could calculate ΔG°_{cell} by first determining ΔG°_f's from Appendix C for each half-reaction. Here we will show the use of Table 13.1 and Equation 13.4. Table 13.1 gives $\mathscr{E}^\circ_{cath} = -0.828$ V and $\mathscr{E}^\circ_{anod} = -1.360$ V. By Equation 13.4 this gives $\Delta G^\circ_{cath} = 38.2$ kcal and $\Delta G^\circ_{anod} = 62.7$ kcal. Thus, $\Delta G^\circ_{cell} = 100.9$ kcal, and $\mathscr{E}^\circ_{cell} = (-100.9$ kcal)/ $[2\,mol(23.0605$ kcal/mol V$)] = -2.188$ V.

To determine the charge that passes through, we first need to determine the mols of $Cl_2(g)$ that occupy 200. L at STP. Equation 9.13 gives 8.93 mol $Cl_2(g)$ formed. Now, the anode half-reaction shows that there are two mols of e^- for every mol of $Cl_2(g)$ that is formed. The amount of charge, then, is given by

$$\text{amount of charge} = 8.93 \,\text{mol}\, Cl_2 \left(\frac{2\,\text{mol}\,e^-}{1\,\text{mol}\,Cl_2}\right)\left(\frac{96{,}485\,C}{1\,\text{mol}\,e^-}\right) = 1.72 \times 10^6\,C$$

47. The half-reactions shown in the text answer key for Exercise 46 reveal that two mols of e^- pass for every mole of $Cu(s)$ that is formed. Equation 13.19 gives $Q = It = (0.145$ C/s)(2700. s) = 391.5 C. The mass of copper that will plate out is, therefore,

$$g(Cu) = 391.5\,C\left(\frac{1\,\text{mol}\,e^-}{96{,}485\,C}\right)\left(\frac{1\,\text{mol}\,Cu}{2\,\text{mol}\,e^-}\right)\left(\frac{63.546\,g}{1\,\text{mol}\,Cu}\right) = 0.129\,g\,Cu$$

49. The reaction that describes this process is $2NaCl(aq) + 2H_2O(l) \rightarrow 2NaOH(aq) + Cl_2(g) + H_2(g)$, where we can envision the $Na^+(aq)$ present in the brine being reduced to Na at the cathode. The Na then turns quickly around and reduces the water. Therefore, there will be one mol of electrons transferred for every mole of NaCl. Conversely, we can say that there are two mols of electrons for every two moles of NaCl and, therefore, NaOH from the balanced equation. Perhaps, it is easier to see that there are two moles of electrons transferred by looking at the Cl half-reaction: $2Cl^- \rightarrow Cl_2 + 2e^-$. Either way, the current is

$$\text{current} = 200. \times 10^3 \text{ kg NaOH} \left(\frac{1000 \text{ g}}{1 \text{ kg}} \right) \left(\frac{1 \text{ mol}}{39.9971 \text{ g}} \right) \left(\frac{2 \text{ mol e}^-}{2 \text{ mol NaOH}} \right) \left(\frac{96,485 \text{ C}}{1 \text{ mol e}^-} \right) \left(\frac{1}{86400 \text{ s}} \right)$$

$$\approx 5.58 \times 10^6 \text{ A}$$

51. In the text we are told that $Fe^{2+}(aq)$ will be rapidly oxidized to $Fe^{3+}(aq)$ by O_2 diffusing into the electrolyte. (The electrolyte is rainwater with a few ions in it.) To form $Fe(OH)_3(s)$, the overall reaction must contain at least Fe^{2+}, OH^-, and O_2 as reactants and $Fe(OH)_3$ as the product. By analogy with Equation 13.20 $H_2O(l)$ needs to be included as a reactant in the overall equation, as is often the case for aqueous redox chemistry. Thus, we expect $Fe^{2+}(aq) + O_2(g) + H_2O(l) + OH^-(aq) \rightarrow Fe(OH)_3(s)$ as the (unbalanced) overall reaction. Additionally, the cathode reaction will be the same as it was in Equation 13.20. We can deduce the anode reaction, as in Exercises 34 and 35, from these two equations to get

$$O_2(g) + 2H_2O(l) + 4e^- \rightarrow 4OH^-(aq) \qquad \text{Cathode}$$
$$[Fe^{2+}(aq) + 3OH^-(aq) \rightarrow Fe(OH)_3(s) + e^-] \times 4 \qquad \text{Anode}$$

$$4Fe^{2+}(aq) + O_2(g) + 2H_2O(l) + 8OH^-(aq) \rightarrow 4Fe(OH)_3(s)$$

As we have shown throughout this chapter, there are several ways to calculate $\mathscr{E}^\circ_{cell}$. The easiest way is to use ΔG°_f values from Appendix C for the overall equation. In light of this chapter, however, we will use a slightly longer, more illustrative approach. For the cathode, Table 13.1 gives $\mathscr{E}^\circ_{cath} = 0.401$ V and Equation 13.4 gives $\Delta G^\circ_{cath} = -4$ mol(23.0605 kcal/mol V)(0.401 V) $= -37.0$ kcal. Since there is no listing for the anode reaction in Table 13.1, we have to use Appendix C. This gives $\Delta G^\circ_{anod} = -34.9$ kcal and $\mathscr{E}^\circ_{anod} = 1.51$ V. So, $\Delta G^\circ_{cath} + 4\Delta G^\circ_{anod} = -176.6$ kcal. Lastly, Equation 13.4 gives $\mathscr{E}^\circ_{cell} = 176.6$ kcal/ [4 mol(23.0605 kcal/mol V)] $= 1.91$ V. We could also have used Equation 13.15, keeping in mind that we have to write the anode reaction as a reduction, by convention. This switches the sign of $\mathscr{E}^\circ_{anod}$ and leads to $\mathscr{E}^\circ_{cell} = 0.401$ V $- (-1.51$ V) $= 1.91$ V.

14

States of Matter
and Intermolecular Forces

Your Chapter 14 GOALS:

- Qualitatively compare potential energy curves for the interaction of different atoms
- Construct heating curves from enthalpy and heat capacity data
- Calculate entropy changes at phase transitions
- Sketch the dependence of free energy on temperature for water
- Use free energy expressions to calculate equilibrium vapor pressures
- Define the difference between *equilibrium* vapor pressure and *any* vapor pressure
- Sketch phase diagrams and identify the key points present in them
- Explain what occurs when water boils
- Identify the attractive intermolecular forces
- Predict which intermolecular forces are at work in various substances
- Calculate the potential energy for the various intermolecular interactions, all of which are all based on the inverse-power law $V(R) \propto 1/R^s$
- Use intermolecular forces to explain physical properties
- Calculate repulsive interactions with the Born-Mayer repulsion
- Describe the patterns seen in crystal formation
- Use Bragg's Law to describe X-ray diffraction and to predict the spacing present in crystal lattices
- Use band theory, an extension of MO theory, to rationalize the conductivity of metals and the non-conducting nature of insulators
- Explain lattice energy as arising from the interactions described by the ionic model
- Use competing enthalpy and entropy factors to discuss solubility
- Combine free energy principles with the properties of dilute solutions to derive Raoult's Law and expressions for boiling point elevation and depression
- Define osmotic pressure

Chapter 14 KEY EQUATIONS:

- 14.3, 14.5, 14.6, 14.8, 14.10, 14.14, 14.16, 14.18, 14.19, 14.24, 14.25, 14.26, 14.27, 14.28

Overview

The role of k_BT in the occupation of vibrational levels is examined in Exercise 21 of Chapter 8. In Section 14.1 we again examine k_BT, but this time we focus on its role as the origin of phase transitions. As T increases the average thermal energy, k_BT, that is available to a system increases. This added energy creates a greater degree of random motion, eventually breaking intermolecular or interatomic "bonds" when the motion becomes sufficiently severe. Only those substances with large intermolecular or interatomic forces are able to remain solids at high T's. The usual phase change from $(s) \rightarrow (l) \rightarrow (g)$ is tracked, typically at 1 atm, in a heating curve, the subject of Section 14.2. Note that we say usual because, for example, at 1 atm CO_2 sublimes, a fact that should be evident from your phase diagram for Exercise 3. When solving Exercise 20, among others, be sure that you remember ΔH°_{fus} is for the melting process, so $\Delta V_{fusion} = V_{liquid} - V_{solid}$, yielding $\Delta V_{fusion} < 0$. Exercises 1 through 5 closely follow Examples 14.1 and 14.2, and it is important to see that the calculations are different for the regions *between* the phase changes as opposed to those *at* the phase changes. Pay special attention to the calculation of ΔG for Exercises 1 and 5. This is more difficult than previous ΔG calculations because T is not constant during the process. To compare Exercise 4 to Example 10.6, note that $\Delta H^\circ = -101.8$ kcal/mol Fe for the thermite reaction.

Exercises 6 through 8 all follow Example 14.3, and the key to these Exercises is that $\Delta G° = 0$ at any phase transition. This fact is true, then, at *any* point along any phase boundary on a phase diagram. Our usual assumption that $\Delta H°$ and $\Delta S°$ are independent of temperature allows the use of Appendix C for their calculation, but it also leads to a small error. Note in Exercise 9 that the derived ΔS formula should not surprise you as it bears a strong resemblance to Equations 12.6 and 13.6. Exercises 9(b) and 10 through 14 all require the use of Equation 14.8. Table 14.1 is the source for the necessary $\Delta H°$ and T_b/T_f data. Exercise 10 can also be solved via Equation 14.9 by obtaining $\Delta S°_{vap}$ from Equation 14.3. For Exercises 15 and 16 note that the general headings of Table 14.1 don't all apply to CO_2. For example, the value listed under $\Delta H°_{fus}$ is actually $\Delta H°_{sub}$ for CO_2. The derivation in Exercise 18 is identical to the one in Exercise 16 of Chapter 12 if T_0 is replaced by T_b. Note that the final form you derive will be the original Clausius-Clapeyron equation with an additional term on the right hand side. Exercise 19 requires estimation of only one value, P_3 for Na. See the solution for the reason.

Using Equation 14.3, Exercise 24 indicates that the solution with the highest degree of order will have the largest entropy increase upon vaporization. As a means of comparison, the organization of water is greater than any of the compounds that are considered in this Exercise, and Equation 14.3 gives $\Delta S°_{vap}(H_2O) = 26.0$ cal/K mol. Water has such a large amount of organization that it exceeds the average value of 22 cal/K mol of Trouton's Rule by 4 cal/K mol, and it is no coincidence that H_2O and CH_3OH *both* exhibit hydrogen bonding and *both* have $\Delta S°_{vap}$'s > 22 cal/K mol. Exercises 25 through 34 are all related and explore the calculation of the various intermolecular forces of Section 14.4, the location of the required equations. Only Exercise 30, however, uses Equation 14.18. After converting the well depth to eV/molecule, you ought to be able to derive the following equations in the two unknowns A and B:
i) $A = 0.01874$ eV/e$^{-(4.40\ Å)B}$ and ii) $ABe^{-(4.40\ Å)B} = 0.05393$ eV/Å.

The Bragg's Law Exercises require that you understand the difference between the angle of incidence θ and the angle of the diffraction ring with respect to the X-ray diffraction beam 2θ. For Exercise 36 $\theta_n < 90°$ only if $n = 1$ or 2, and the $2\theta_n$ angles are twice this. Exercise 37 deals with the same concerns over θ_n, where $\theta_2 < 90°$ and, therefore, $2\theta < 180°$. For Exercise 38 use the volume of a sphere to calculate the portion of the cell volume that an atom occupies, being sure to multiply by the number of atoms that are present in the unit cell. This occupied volume must then be divided by the available volume, which is the cell volume a^3. The band theory of solids is essentially LCAO-MO theory scaled up to *many* atoms, as shown in Figure 14.19. Use PiB nodal structure arguments to sketch the MO's of Li_5 for Exercise 40. To complete the sketch accurately, however, see the solution to Exercise 27 and 28 of Chapter 7 for H_3, which shows that sometimes AO's are excluded from the final MO's. The case is a little different here because there are 5 atoms, but try eliminating AO's to get MO's that match your PiB nodal structure predictions. As you solve Exercise 46, note that the bcc and fcc units cells have nearest neighbors that lie along the interior diagonal and face diagonal, respectively. Therefore, Exercise 46 requires a bit more geometry work than Exercise 45. Exercise 46 requires use of the Pythagorean formula with the face diagonal, followed by its use again with the interior diagonal.

Exercise 48 puts a quantitative spin on the qualitative meaning of Table 12.6 or even the solubility rules of Chapter 5. Since chlorides are "soluble", we predict that dissolving $MgCl_2$ in water is a spontaneous process. Hydroxides are listed in Table 12.6, a fact that by itself indicates that there is a good chance the dissolving of $Mg(OH)_2$ in water is not spontaneous. We verify these predictions quantitatively with Appendix C and can also be more qualitative in our reasoning for it by appealing to solute/solvent interactions. Indeed, we have come a long way from the simple designation of Chapter 5 of compounds being "soluble" or "insoluble". For part (c) of Exercise 52, you will need to make the approximation that $n_A + n_B \approx n_A$ since $n_B \ll n_A$ at some point along the way. Remember for Exercises 53 and 54 that Equations 14.25 and 14.26 contain ΔT and that ΔT must be added or subtracted appropriately from the normal phase transition temperature. Exercises 56 and 60 reinforce the rule that colligative properties depend on the moles of solute particles, not the moles of solute. Also, when the ratio of Equation 14.27 is taken for Exercise 56, K_f and the solute and solvent masses cancel.

14 States of Matter and Intermolecular Forces

1. a. Similar to Example 14.1 there are three different stages to consider. For the first stage, using $q = n\overline{C}_p\Delta T$ and the density of water,

$$q_1 = 1.00 \text{ L}\left(\frac{1000 \text{ cm}^3}{1 \text{ L}}\right)\left(\frac{1.00 \text{ g}}{1 \text{ cm}^3}\right)\left(\frac{1 \text{ mol}}{18.0152 \text{ g}}\right)\left(\frac{18.0 \text{ cal}}{\text{K mol}}\right)\left(\frac{-25 \text{ K}}{1}\right) = -24980 \text{ cal}$$

At the freezing point, $q_2 = n\Delta H^\circ_{fus} = 55.51$ mols(1.436 kcal/mol) $= -79.71$ kcal. We add the negative sign because heat is released. In the final step the ice cools, and we use the equation from the first stage, except that the molar heat capacity is that for the solid. So,

$$q_3 = 55.51 \text{ mol}\left(\frac{9.0 \text{ cal}}{\text{K mol}}\right)\left(\frac{-15 \text{ K}}{1}\right) = -7490 \text{ cal}$$

Thus, $q_{tot} = q_1 + q_2 + q_3 = q_p = \Delta H = -112.2$ kcal.

b. Equation 14.2 gives the entropy change for stages one and three, using the appropriate molar heat capacity value. For stage one,

$$\Delta S_1 = 1.00 \text{ L}\left(\frac{1000 \text{ cm}^3}{1 \text{ L}}\right)\left(\frac{1.00 \text{ g}}{1 \text{ cm}^3}\right)\left(\frac{1 \text{ mol}}{18.0152 \text{ g}}\right)\left(\frac{18.0 \text{ cal}}{\text{K mol}}\right)\ln\left(\frac{273}{298}\right) = \frac{-87.55 \text{ cal}}{\text{K}}$$

At the freezing point (or boiling point) note that $\Delta S = q/T$, where q (calculated above) is the total heat that is lost as the phase change occurs. Thus, $\Delta S_2 = -79.71$ kcal/273 K $= -0.292$ kcal/K. For the solid we use the appropriate molar heat capacity and (258/273) as the argument of ln to get $\Delta S_3 = -28.23$ cal/K. So, $\Delta S_{tot} = \Delta S_1 + \Delta S_2 + \Delta S_3 = -0.408$ kcal/K.

The ΔG calculation is different from others in previous chapters. The definition of free energy is $G = H - TS$. Thus, $\Delta G = \Delta H - \Delta(TS)$. Equation 11.24 can't be used since the temperature is not constant. Now, since S and T are state functions $\Delta G = \Delta H - [S_f T_f - S_i T_i]$. f = final, and i = initial. There are 55.51 mols, and from Appendix C for $H_2O(l)$ $S_{initial} = 55.51$ mols(16.71 cal/K mol) $= 0.928$ kcal/K. Recall that we CAN speak about the 'entropy' of something in *absolute* terms because of the Third Law. Now, $\Delta S_{tot} = -0.408$ kcal/K, which leads to $S_{final} = 0.520$ kcal/K. So, $\Delta G = -112.2$ kcal $- [(0.520$ kcal/K)258 K $- (0.928$ kcal/K)298 K] $= 30.2$ kcal.

3. For the range of 194.7 K (-78.5 °C) to 258.2 K (-15 °C), $\Delta T = 63.5$ K. 194.7 K (-78.5 °C) is the sublimation temperature. Thus, there is no change in temperature until all of the $CO_2(s)$ sublimes. For the heating of one mol, Table 14.1 and its footnote for CO_2 immediately show that 5.9 kcal are needed. The second stage heats the gas. Thus, $q_2 = 1$ mol(8.9 cal/K mol)(63.5 K) $= 565$ cal. These values appear in the figure below, and as in Figure 14.3, the H° values come from the standard heat of formation of $CO_2(g)$ at 298.15 K (-94 kcal/mol). This curve is different from water and most substances at 1 atm because there is only a $(s) \to (g)$ transformation, not $(s) \to (l) \to (g)$.

Whereas the first part assumed one mole, 1.00 kg is (1.00 $\times$ 10^3 g CO_2/(44.010 g/mol) $\approx$) 22.7 mols. So, $q_1 = 134$ kcal and $q_2 = 12.8$ kcal. Thus, $q_{tot} = 147$ kcal. Any amount of heat greater than this will result in a temperature higher than -15 °C.

14-4

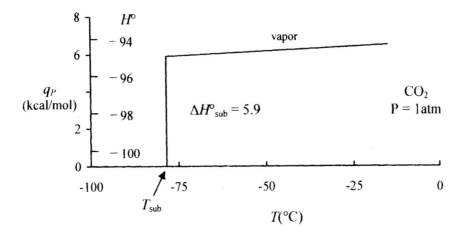

5. a. Heating the solid from 298 K to 370.9 K, the melting point, requires $n\bar{C}_p(s)\Delta T$ kcal. (Note how low this melting point is. This is typical of the group IA metals.) Thus,

$$q_1 = 100.\,kg\left(\frac{1000\,g}{1\,kg}\right)\left(\frac{1\,mol}{22.9898\,g}\right)\left(\frac{6.7\,kcal}{K\,mol}\right)\left(\frac{72.9\,K}{1}\right) = 2125\,kcal$$

At the melting point the amount of heat required to complete the phase change is $q_2 = n\Delta H^\circ_{fus} = 4349.8$ mol(0.62 kcal/mol) = 2697 kcal. So, $q_{tot} = q_1 + q_2 = q_p = \Delta H = 4822$ kcal.

b. For stage 1 Equation 14.2 gives

$$\Delta S_1 = 4349.8\,mol\left(\frac{6.7\,cal}{K\,mol}\right)\ln\left(\frac{370.9\,K}{298\,K}\right) = \frac{6378\,cal}{K}$$

At the boiling point (or freezing point) we note that $\Delta S = q/T$, where q is the total heat that is added as the phase change occurs. We calculated this above. Thus, $\Delta S_2 = 2697$ kcal/370.9 K = 7.27 kcal/K. So, $\Delta S_{tot} = 13.65$ kcal/K.

The ΔG calculation is different from others done in previous chapters. The definition of free energy is $G = H - TS$. Thus, $\Delta G = \Delta H - \Delta(TS)$. Equation 11.24 can't be used since the temperature is not constant. Now, since S and T are state functions $\Delta G = \Delta H - [S_f T_f - S_i T_i]$. f = final, and i = initial. We have 4349.8 mols, and from Appendix C for Na(s) $S_{initial} = 4349.8$ mols(12.24 cal/K mol) = 53.24 kcal/K. Recall that we CAN speak about the 'entropy' of something in absolute terms because of the Third Law. Now, $\Delta S_{tot} = 13.65$ kcal/K, which leads to $S_{final} = 66.89$ kcal/K. Thus, $\Delta G = 4822$ kcal – [(66.89 kcal/K)370.9 K – (53.24 kcal/K)298 K] = – 4120 kcal.

7. As in Example 14.3, ΔG° is zero at a transition temperature. So, $T_{trans} = \Delta H^\circ/\Delta S^\circ$. For the phase change equation $CH_3CH_2OH(l) \rightarrow CH_3CH_2OH(g)$ Appendix C gives $\Delta H^\circ_{298} = 10.18$ kcal and $\Delta S^\circ_{298} = 29.1$ cal/K. Thus, $T_b = 10.18$ kcal/(0.0291 kcal/K) = 349.8 K ≈ 76.7 °C. Now, Table 14.1 gives $T_b = 351.4$ K ≈ 78.3 °C. The 1.6 °C difference is due to the fact that ΔH and ΔS depend slightly on temperature. The calculations, however, are greatly simplified when we assume that they do not.

9. a. The assumptions here are those made in Example 14.3. Thus, $T_{ta} \approx T_b = \Delta H_{vap}/\Delta S_{vap}$. 510 Torr is not a standard pressure and the superscript zeroes are not used. However, given $\Delta H \approx \Delta H^\circ$, the ΔH°_{vap} value in Table 14.1 can be used. To develop an equation to calculate ΔS_{vap}, recall Equation 11.11, which can be regarded as a correction factor to

ΔS°_{vap}, yielding $\Delta S_{vap} = \Delta S^{\circ}_{vap} + nR\ln(P^{\circ}/P)$. We have used Boyle's Law, where standard pressure is regarded as the initial pressure, to rewrite the volume ratio. With the pressure expressed in atmospheres, $\Delta S_{vap} = \Delta S^{\circ}_{vap} - nR\ln P$, where P is the (non-standard) pressure of the gas. ΔS°_{vap} is calculated from Equation 14.3. This gives (9.716 kcal/mol)/373.15 K = 0.02604 kcal/K mol. So, converting the pressure to atmospheres,

$$\Delta S_{vap} = \frac{0.02604\,\text{kcal}}{\text{K}} - 1\,\text{mol}\left(\frac{0.0019872\,\text{kcal}}{\text{K mol}}\right)\ln\left(\frac{510}{760}\right) = \frac{0.02683\,\text{kcal}}{\text{K}}$$

Note that we dropped the mol^{-1} from ΔS°_{vap}, since the problem is for one mol. Lastly, $T_b = (9.716\,\text{kcal/mol})/(0.02683\,\text{kcal/K mol}) = 362.1\,\text{K} \approx 89.0\,°\text{C}$.

b. Substituting the values from Table 14.1 into Equation 14.8 gives

$$\ln\left(\frac{\dfrac{510\,\text{Torr}}{760\,\text{Torr}/1\,\text{atm}}}{1\,\text{atm}}\right) = \frac{-\dfrac{9.716\,\text{kcal}}{\text{mol}}}{\dfrac{0.0019872\,\text{kcal}}{\text{K mol}}}\left[\frac{1}{T_b} - \frac{1}{373.15\,\text{K}}\right]$$

This is solved to give $T_b = 362.1\,\text{K} = 89.0\,°\text{C}$.

11. a. Using Equation 14.8 to obtain the equilibrium vapor pressure of water P_{eq} at 305 K, with the reference point T_b, gives

$$\ln\left(\frac{P_{eq}}{1}\right) = \frac{-\dfrac{9.716\,\text{kcal}}{\text{mol}}}{\dfrac{0.0019872\,\text{kcal}}{\text{K mol}}}\left[\frac{1}{305\,\text{K}} - \frac{1}{373.15\,\text{K}}\right] \quad \Rightarrow \quad P_{eq} = 0.0535\,\text{atm}$$

If the relative humidity is 70.0 %, the vapor pressure (or partial pressure) of water vapor P_{vap} is only 0.700 times P_{eq}. This gives

$$\frac{P_{vap}}{0.0535\,\text{atm}} \times 100\% = 70.0\% \quad \Rightarrow \quad P_{vap} = 0.0375\,\text{atm} \approx 28.5\,\text{Torr}$$

b. Assuming the water temperature is the same as that of the air, the water in the pool will not be at equilibrium with its vapor since $P_{vap} < P_{eq}$. The free energy change is given by Equation 14.7. So,

$$\Delta G_{vap} = \frac{0.0019872\,\text{kcal}}{\text{K mol}}(305\,\text{K})\ln\left(\frac{0.0375\,\text{atm}}{0.0535\,\text{atm}}\right) = \frac{-0.215\,\text{kcal}}{\text{mol}}$$

Note that the sign is negative, indicating that the process of evaporation is spontaneous at these conditions. This is clearly in line with our experience.

c. Loosely stated, condensation is spontaneous at the dew point because the air can't "hold" any more water vapor. To be more quantitative, note that as T gets smaller P_{eq} gets smaller via Equation 14.6. When P_{vap} becomes greater than or equal to P_{eq}, H_2O "drops out" of the air as dew. For this Exercise this is the temperature at which $P_{eq} \leq 0.0375$ atm. Solving Equation 14.8 gives $T_{dew} = 298.4\,\text{K} = 25.3\,°\text{C}$ via

$$\ln\left(\frac{0.0375\,\text{atm}}{1\,\text{atm}}\right) = \frac{-\dfrac{9.716\,\text{kcal}}{\text{mol}}}{\dfrac{0.0019872\,\text{kcal}}{\text{K mol}}}\left[\frac{1}{T_{dew}} - \frac{1}{373.15\,\text{K}}\right]$$

13. a. As discussed in the text, heating ice at pressures below the triple point pressure (0.0060 atm ≈ 4.6 Torr) results in sublimation, meaning no $(s) \rightarrow (l)$ transition.

   b. Replacing $\Delta H°_{vap}$ with $\Delta H°_{sub}$ in Equation 14.8 and using (T_3, P_3) as the reference point, as mentioned in the hint of Exercise 12, gives $T_{sub} = 251.1$ K $= -22.1$ °C:

$$\ln\left(\frac{(0.75/760)\,\text{atm}}{0.0060\,\text{atm}}\right) = \frac{11.152\,\dfrac{\text{kcal}}{\text{mol}}}{0.0019872\,\dfrac{\text{kcal}}{\text{K mol}}}\left[\frac{1}{T_{sub}} - \frac{1}{273.16\,\text{K}}\right]$$

15. For use with Equation 14.8 (T_3, P_3) provides one (P, T) pair. For the other (P, T) pair recall that CO_2 sublimation occurs at 1 atm of pressure and, from Table 14.1, 194.7 K. Although this number is found under the heading of T_f in Table 14.1, the footnote tells us that this is T_{sub} in the case of CO_2. Therefore,

$$\ln\left(\frac{5.15\,\text{atm}}{1\,\text{atm}}\right) = \frac{-\Delta H°_{sub}}{0.0019872\,\dfrac{\text{kcal}}{\text{K mol}}}\left[\frac{1}{216.6\,\text{K}} - \frac{1}{194.7\,\text{K}}\right]$$

Thus, $\Delta H°_{sub} = 6.27$ kcal/mol. Taking 5.9 kcal/mol in Table 14.1 to be the accepted value, this is a percent error of 6.3 %.

17. Using Table 14.1, the following table is constructed:

element	T_c/T_b	compound	T_c/T_b
Ne	1.64	HF	1.57
Ar	1.73	HCl	1.73
Kr	1.75	HBr	1.76
Xe	1.75	HI	1.78

The average value of T_c/T_b is 13.71/8 = 1.71. For water $T_c = 1.71(T_b) = 1.71(373.15$ K$) = 638$ K. Likewise, $T_c = 317$ K and 275 K for PH_3 and SiH_4, respectively. The percent error for water is $(638 - 647.1)/647.1 = -1.41$ %. Likewise, PH_3 and SiH_4 have percent errors of -2.31 % and 1.97 %, respectively.

19. ALL of the phase diagrams for this problem have the shape of the phase diagram shown for water in Figure 14.8. Simply identify the four key points for each element. From Table 14.1 we can develop the following table:

	(T, P)			
	normal freezing point	normal boiling point	triple point	critical point
HCl	(159.0, 1)	(188, 1)	(159.3, 0.14)	(324.7, 82.0)
Hg	(234.3, 1)	(629.9, 1)	(234.3, 4×10^{-9})	(1765, 1490.)
NH_3	(195.4, 1)	(239.8, 1)	(195.4, 0.060)	(405.5, 112.0)
Na	(370.9, 1)	(1156, 1)	(370.7, 3×10^{-9})	(2083, 253)
O_2	(54.4, 1)	(90.2, 1)	(54.4, 0.0015)	(154.6, 49.8)

Only P_3 for Na must be calculated. Always be aware that tables of data can contain *calculated* data. This is especially true for extreme conditions. From Equation 14.8

$$\ln\left(\frac{P_3}{1}\right) = \frac{-\dfrac{21.4\,\text{kcal}}{\text{mol}}}{0.0019872\,\dfrac{\text{kcal}}{\text{K mol}}}\left(\frac{1}{370.7\,\text{K}} - \frac{1}{1156\,\text{K}}\right) \Rightarrow P_3 = 3\times10^{-9}\,\text{atm}$$

Note that the minuscule pressures for Hg and Na require a log scale to be used.

21. Following the hint, Equation 14.12 (I) becomes $dq = dE + PdV$ and Equation 14.12 (II) becomes $dq = TdS$. Equating the dq's gives $dE + PdV = TdS$ (relation 1). The definition of H is Equation 10.22. In differential form this is $dH = dE + PdV + VdP$ (relation 2), using the rule for the differential of a product. The definition of G is Equation 11.22. In differential form this is $dG = dH - TdS - SdT$ (relation 3). Substituting relations 1 and 2 into relation 3 gives $dG = dE + PdV + VdP - dE - PdV - SdT$. This reduces to $dG = VdP - SdT$. This can apply to any phase at equilibrium, giving $dG_1 = V_1dP - S_1dT$ and $dG_2 = V_2dP - S_2dT$. 1 and 2 are different phases. At equilibrium, $dG_1 = dG_2$, and $V_1dP - S_1dT = V_2dP - S_2dT$. So, $(S_2 - S_1)dT = (V_2 - V_1)dP$, or $dP/dT = \Delta S/\Delta V$. Finally, at equilibrium $\Delta G = 0$ and $\Delta H = T\Delta S$. This gives the Clausius equation: $dP/dT = \Delta H/T\Delta V$.

23. From Trouton's rule $\Delta H^\circ_{vap} = (22\,\text{cal/K mol})T_b$. Inserting this into the Clausius-Clapeyron equation with the normal boiling point as the reference gives

$$\ln\left(\frac{P}{1}\right) = \frac{-(0.022\,\text{kcal/K mol})T_b}{0.0019872\,\text{kcal/K mol}}\left(\frac{1}{T} - \frac{1}{T_b}\right)$$

Simplifying, $\ln P = -11T_b[(1/T) - (1/T_b)]$, or $\ln P = 11(1 - T_b/T)$. Using Table 14.1 to find $T_b = 329.2\,\text{K}$ and the given $T = 298.15\,\text{K}$, the approximate equation yields $P = 0.318\,\text{atm} \approx 242$ Torr. The Clausius-Clapeyron equation yields

$$\ln\left(\frac{P}{1}\right) = \frac{-6.96\,\text{kcal/mol}}{0.0019872\,\text{kcal/K mol}}\left(\frac{1}{298.15\,\text{K}} - \frac{1}{329.2\,\text{K}}\right) \Rightarrow P = 0.330\,\text{atm} \approx 251\,\text{Torr.}$$

The Clausius-Clapeyron equation should be more accurate. It is not in this case, but it usually is since it uses the specific ΔH°_{vap} for the substance of interest.

25. Given in Section 14.4, the magnitude of the ion-dipole interaction potential is $V(R) = ez_A\mu_B/R^2$. For the Na^+/dipole interaction to be attractive ($V(R) < 0$), the Na^+ must be positioned near the oxygen end of the water molecule, the negative end of the dipole. Adding a negative sign to indicate attraction,

$$V(R) = \frac{-4.8032\times10^{-10}\,[\text{erg cm}]^{1/2}(1)(1.85\times10^{-18}\,\text{esu cm})}{(2.35\times10^{-8}\,\text{cm})^2} = \frac{-1.61\times10^{-12}\,\text{erg}}{\text{molecule}}$$

(We can also think of the unit as erg per interaction.) The units do, in fact, work out: (erg$^{1/2}$ cm$^{1/2}$ erg$^{1/2}$ cm$^{1/2}$ cm)/cm^2 = erg. Using the two given conversion factors gives -23.2 kcal/mol ≈ -97.0 kJ/mol. You should be able to figure out from where these conversions come. For the Cl^-/dipole interaction to be attractive ($V(R) < 0$), the Cl^- must be positioned near the hydrogen atoms, the positive end of the dipole. The calculation is the same as above, except that R now equals 3.00×10^{-8} cm. Thus, $V(R) = -9.87 \times 10^{-13}$ erg/molecule ≈ -14.2 kcal/mol ≈ -59.4 kJ/mol. The Na^+ produces a more negative potential energy, meaning that it is the more strongly solvated. The reason for this is that the equation contains an inverse-square relation with respect to R. The Cl^- ionic radius is bigger, increasing the size of R and decreasing the magnitude of $V(R)$.

27. The equation for the magnitude of the dipole-dipole interaction between two polar molecules is $V(R) = 2\mu_A\mu_B / R^3$. For NH_3 and H_2O, $R = 1.73 \times 10^{-8}$ cm $+ 1.50 \times 10^{-8}$ cm $= 3.23 \times 10^{-8}$ cm. Thus,

$$V(R) = \frac{-2(1.49 \times 10^{-18} \text{esu cm})(1.85 \times 10^{-18} \text{esu cm})}{(3.23 \times 10^{-8} \text{cm})^3} = \frac{-1.64 \times 10^{-13} \text{ erg}}{\text{molecule}} \approx \frac{-2.35 \text{ kcal}}{\text{mol}}$$

This is equal to -9.85 kJ/mol. A related unit analysis was done in Exercise 25. The negative sign shows an attractive interaction. This is slightly smaller than the value for two water molecules, and the difference is due to a larger intermolecular distance and a smaller dipole moment for NH_3.

29. All of the calculations for this problem are done in the identical manner. (a) and (b) are shown in detail. It is often useful to predict the results before solving a problem, but this is difficult here since there are three factors, the polarizability, the IE's, and the r_{vdW}'s, that influence the outcome. To simplify the math, note that for (a) through (e), we can rewrite the London dispersion coefficient as $C = (\frac{3}{4})\alpha^2 I$.

a. For the Ar---Ar interaction

$$C = \frac{3}{4}\alpha^2 I = \frac{3}{4}\left[(1.64 \text{ Å}^3)^2(15.76 \text{ eV})\right] = \frac{31.8 \text{ eV Å}^6}{\text{molecule}}$$

Next, the London dispersion energy comes from $V(R) = -C/R^6$:

$$V(R) = \frac{-\dfrac{31.8 \text{ eV Å}^6}{\text{molecule}}}{(3.76 \text{ Å})^6}$$

$$= \frac{-0.01125 \text{ eV}}{\text{molecule}}\left(\frac{1.6022 \times 10^{-19} \text{ J}}{\text{eV}}\right)\left(\frac{6.0221 \times 10^{23} \text{ molecules}}{\text{mol}}\right)\left(\frac{1 \text{ kJ}}{1000 \text{ J}}\right)$$

$$= \frac{-1.09 \text{ kJ}}{\text{mol}} \approx \frac{-0.260 \text{ kcal}}{\text{mol}}$$

The interaction has a negative potential energy; this force is always attractive.

b. For the H_2O---H_2O interaction

$$C = \frac{3}{4}\alpha^2 I = \frac{3}{4}\left[(1.45 \text{ Å}^3)^2(12.6 \text{ eV})\right] = \frac{19.9 \text{ eV Å}^6}{\text{molecule}}$$

Next, the London dispersion energy comes from $V(R) = -C/R^6$:

$$V(R) = \frac{-\dfrac{19.9 \text{ eV Å}^6}{\text{molecule}}}{(3.00 \text{ Å})^6}$$

$$= \frac{-0.0273 \text{ eV}}{\text{molecule}}\left(\frac{1.6022 \times 10^{-19} \text{ J}}{\text{eV}}\right)\left(\frac{6.0221 \times 10^{23} \text{ molecules}}{\text{mol}}\right)\left(\frac{1 \text{ kJ}}{1000 \text{ J}}\right)$$

$$= \frac{-2.63 \text{ kJ}}{\text{mol}} \approx \frac{-0.629 \text{ kcal}}{\text{mol}}$$

See the text key for (c) through (g).

31. The London dispersion value is -0.63 kcal/mol. The dipole-dipole value is -3.65 kcal/mol, and their sum is -4.28 kcal/mol. This is close to the hydrogen bond strength, which is the third attractive force between water molecules in addition to the LD and DD forces. Lastly,

we have omitted the Born-Mayer repulsion. Of course, a repulsive force will raise the potential energy and lessen the attraction between two water molecules.

33. If you look back at the previous Exercises, the only force not yet calculated is the LD force for H_2O---CH_4. 2 sample calculations were shown for this in the solution to Exercise 29. CH_4 has no dipole moment by symmetry. Refer to the text key for the table.

Water is a liquid while CH_4 is a gas at NTP because of the much greater attraction between the water molecules, evidenced by the values of the sum of the LD and DD energies. (Of course, water hydrogen bonds, too.) In the case of polar molecules such as H_2O and NH_3, "like dissolves like" mainly because of the attraction between dipoles. In the case of nonpolar molecules, it is mainly an increase in entropy that results in solution formation. Therefore, the energetic price of breaking bonds, as the "holes" in Figure 14.33 are made, is offset by the new bonds that are made and/or the entropic benefits. Any slight solubility of a nonpolar molecule, such as CH_4, in H_2O can be explained by the formation of H_2O "clathrates" or "cages" around the nonpolar molecules. This water organization, however, decreases the system's entropy, inhibiting significant solubility of nonpolar molecules in water.

35. Bragg's Law quickly leads to $d = n\lambda/2\sin\theta = 1(1.476$ Å$)/2\sin(42.2/2) = 2.050$ Å. Note that we have to take ½ of the angle 2θ, since θ is the angle of incidence and 2θ is the angle of the diffraction ring with respect to the X-ray beam diffraction. (See Figs. 14.17 and 14.18.) Now, there is a second order ring if θ_2 from Bragg's Law (with $n = 2$) is less than 90°: $\theta_2 = \sin^{-1}[2(1.476$ Å$)/2(2.050$ Å$)] = 46.05°$. Since $2\theta_2 = 92.10° < 180°$, we predict that there will be a second order ring at this angle.

37. The spacing is easily found by solving Bragg's Law for d. This gives $d = n\lambda/2\sin\theta = 1(1.793$ Å$)/2\sin(50.98°/2) = 2.083$ Å. Substituting this value into the given equation gives
$$a = 2.083 \text{ Å}\sqrt{1^2 + 1^2 + 1^2} = 3.608 \text{ Å}$$
Now, Figure 14.15 shows that for a fcc lattice $r = [(2)^{1/2}/4]a$. With a from above $r = 1.276$ Å. Now, to determine if there are any higher order rings, we solve Bragg's Law for θ_2 with $n = 2$. Thus, $\theta_2 = \sin^{-1}[2(1.793$ Å$)/2(2.083$ Å$)] = 59.40°$. Since this angle is less than 90°, 2θ will be less than 180°, and a second order ring should exist.

39. Figure 14.15 shows that there are four atoms per unit cell for the fcc lattice, and each unit cell has a volume a^3. Thus,
$$N_A = \frac{4 \text{ atoms}}{(4.0862 \times 10^{-8} \text{ cm})^3}\left(\frac{1 \text{ cm}^3}{10.501 \text{ g}}\right)\left(\frac{107.868 \text{ amu}}{\text{atom}}\right) = \frac{6.0223 \times 10^{23} \text{ amu}}{\text{g}}$$
Using the unit of g/mol in the last conversion factor yields the more familiar unit atoms/mol.

41. The sketch on the next page. As in s bonding, the number of nodes increases by one for each level, and as the number of atoms increases, the energy of the highest $2s$ level will overlap with the energy of the lowest energy $2p\sigma$ level. This happens because the energy of the highest $2s$ level increases with more atoms due to more nodes, whereas the energy of the lowest $2p\sigma$ level decreases due to an increase in the number of bonding interactions. Additionally, there will be a "mixing" or hybridization of s and $p\sigma$ orbitals since they are close in energy and *of the same symmetry*.

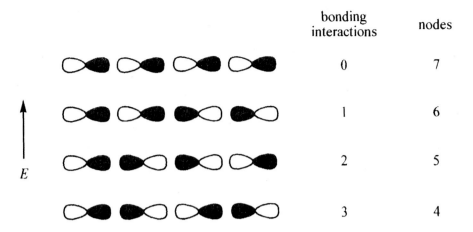

43. As explained in Section 14.5 for Li and Be, there must be overlap between the highest occupied energy level and lowest unoccupied energy level. Electrical conduction will occur always in this lowest unoccupied energy level. This will occur, however, only if the levels overlap or are very closely spaced. Figure 14.22 shows the case for diamond, where the 5 eV energy gap is far too large to allow conduction. Good conductors do not have this large gap. The sketch is very much like Figure 14.20, except that the $2s$ and $2p$ levels are full and that it is the $3s$ and $3p$ levels that overlap, allowing Mg to be a good conductor despite a filled $3s$. Note that Na will look very similar to Li since it has a ½ filled $3s$. Of course, Z changes from the +3 shown in Figure 14.20.

45. Figure 14.15 contains the relevant equation because the three Cl^- ions touch along the face diagonal. The radius for Cl^- is $r_- = [(2)^{1/2}/4](5.64 \text{ Å}) = 1.994 \text{ Å}$. Now, if the cell edge length is made up of two Cl^- radii and one complete Na^+ ion, the diameter of the Na^+ ion must be $(5.64 - 2(1.994))\text{Å} = 1.652 \text{ Å}$. The radius, therefore, is $0.826 \text{ Å} = r_+$. We have retained one more significant figure than is justified.

47. The Born-Haber cycle is below. For $Na(s) + ½ I_2(s) \rightarrow NaI(s)$, $\Delta H°_f = -68.78$ kcal/mol from Appendix C. The other part of the cycle can be determined by several equivalent calculations. To be analogous with Figure 14.28, arrow A represents the value for $\Delta H°_{sub}(Na) + ½\Delta H°_{sub}(I_2) + ½D_0(I_2) = [25.65 + ½ (14.923 + 35.51)]$kcal/mol = 50.87 kcal/mol. The 35.51 kcal value comes from converting the eV D_0 value of Table 8.3. Arrow B represents $IE(Na) - EA(I) = (5.14 - 3.059)$eV = 2.08 eV = 48.0 kcal/mol. This method gives $E_L = [-68.78 - (50.87 + 48.0)]$kcal/mol = -167.7 kcal/mol. Alternatively, since the "top" level of the enthalpy diagram consists of $Na^+(g)$ and $I^-(g)$, one can simply add the $\Delta H°_f$ values for the gas phase ions from Appendix C, 145.55 kcal/mol and -47.1 kcal/mol. This amounts to combining arrows A and B. Thus, $E_L = [-68.78 - (145.55 - 47.1)]$ kcal/mol = -167.2 kcal/mol. Both values are close to the ionic model value of -163 kcal/mol. The two methods yield slightly different values due to the combination of experimental and calculated values that are always present in large tables of data, such as Appendix C.

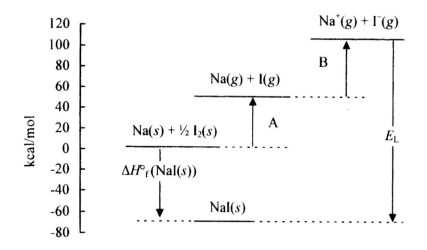

49. The left side of Equation 14.24 can be rewritten as $\ln P_A - \ln X_A$, using the rules for (natural) logarithms. At the solution boiling point $P_A = 1$ atm, and the left side becomes $-\ln X_A$. One of the approximations given below Equation 14.24 allows us to directly convert the left hand side simply to X_B. Now, since $(1/T - 1/T_b) \approx -\Delta T_b/T_b^2$,

$$X_B = \frac{-\Delta H^\circ{}_{vap}}{R}\left(\frac{-\Delta T_b}{T_b^2}\right) \implies \Delta T_b = \frac{X_B R T_b^2}{\Delta H^\circ{}_{vap}}$$

The last given approximation is $m_B \approx 1000 X_B/M_A$. Or, $X_B \approx M_A m_B/1000$. Substitution of this into the above equation gives Equation 14.25.

51. As in Equation 14.22 where $G^\circ{}_A(g) - G^\circ{}_A(l) = \Delta G^\circ{}_{vap}$, here $G^\circ{}_A(l) - G^\circ{}_A(s) = \Delta G^\circ{}_{fus} = -RT\ln X_A$. Recall that fusion refers to the melting process as in Exercise 20. Thus, we have liquid – solid, leading to $\Delta G^\circ{}_{fus}/RT = -\ln X_A$. Equation 11.24 followed by Equation 14.3 then gives

$$-\ln X_A = \frac{(\Delta H^\circ{}_{fus} - T\Delta S^\circ{}_{fus})}{RT} = \frac{\Delta H^\circ{}_{fus}}{RT} - \frac{\Delta S^\circ{}_{fus}}{R}$$

$$= \frac{\Delta H^\circ{}_{fus}}{RT} - \frac{\Delta H^\circ{}_{fus}}{RT_f} = \frac{\Delta H^\circ{}_{fus}}{R}\left[\frac{1}{T} - \frac{1}{T_f}\right]$$

Multiplying both sides by -1 gives the desired equation.

53. Yummy! With a molar mass of 342.299 g/mol, there are 0.2921 mols of sucrose. Thus, Equation 14.25 yields

$$\Delta T_b = K_b m_B = \frac{0.512\,°C\,kg}{mol}\left(\frac{0.2921\,mol}{0.200\,kg}\right) = 0.75\,°C$$

With a normal boiling point of 100.0 °C, this indicates an elevated boiling point of 100.75 °C. (We used the density of water to get 0.200 kg of water.)

55. From Exercise 9 the boiling point for water at 510 Torr is 89.0 °C, or 362.2 K. K_b then becomes

$$K_b = \frac{\dfrac{0.0019872\,kcal}{K\,mol}(362.2\,K)^2\left(\dfrac{18.0152\,g}{mol}\right)}{1000\left(\dfrac{9.716\,kcal}{mol}\right)} = \frac{0.483\,K\,kg}{mol}$$

Now, the increase in boiling point that we need is 11.0 °C (or K) in order to get T_b back to 100.0 °C. The molality of NaCl needed for this is given by Equation 14.27, rewritten for boiling point: $\Delta T_b = iK_b m_B$. Note that this equation contains the van't Hoff factor i. For NaCl there are two moles of ions for every one mol of NaCl that is dissolved, so $i = 2$. Therefore, $m_B = 11.0\ \text{K}/2(0.483\ \text{K kg/mol}) = 11.4$ mol NaCl/kg water. Now, the problem is for 1.00 L, or 1.00 kg, of water. Therefore, 11.4 mol of NaCl or 11.4 mol(58.443 g/mol) = 666 g of NaCl are needed.

57. From text above Example 14.7 the normal melting point of camphor is 178.4 °C and $K_f = 39.7$ °C kg/mol. Now, $\Delta T_f = (178.4 - 171.3)°C = 7.1$ °C. The compound is organic, and therefore not ionic, so Equation 14.25 gives $m_{unk} = 7.1\ °C/39.7\ °C\ \text{kg/mol} = 0.1788$ mol solute/kg solvent. Using the definition of molality, the molar mass of the unknown (M_{unk}) is given by

$$\frac{0.1788\ \text{mol}}{\text{kg}} = \frac{\dfrac{0.0500\ \text{g}}{M_{unk}}}{0.00100\ \text{kg}} \Rightarrow M_{unk} = \frac{280.\ \text{g}}{\text{mol}}$$

59. a. Given h, the osmotic pressure is determined from $\Pi = \rho g h$:

$$\Pi = \frac{980\ \text{kg}}{\text{m}^3}\left(\frac{9.8067\ \text{m}}{\text{s}^2}\right)(0.0510\ \text{m}) = 490.14\ \text{Pa}\left(\frac{1\ \text{atm}}{101{,}325\ \text{Pa}}\right)$$

$$\approx 0.00484\ \text{atm} \approx 3.68\ \text{Torr}$$

b. Using Equation 14.28 in SI units gives (concentration = mols/m^3)

$$490.14\ \text{Pa} = \frac{\dfrac{0.00700\ \text{kg}}{M}}{0.00100\ \text{m}^3}\left(\frac{8.3145\ \text{J}}{\text{K mol}}\right)(298\ \text{K}) \Rightarrow M = \frac{35.39\ \text{kg}}{\text{mol}} \approx \frac{35400\ \text{g}}{\text{mol}}$$

As with empirical formulas, we find

$$n = \frac{\text{mass of polymer}}{\text{mass of monomer}} = \frac{35390\ \text{g}}{62.499\ \text{g}} = 566$$

Therefore, a polymer of PVC consists of a chain of (approximately) 566 monomers, joined end to end.

15

Rates and Mechanisms
of Chemical Reactions

15 Rates and Mechanisms of Chemical Reactions

- Relate the reaction rate to the rate of change of concentration of the reactants and products of a reaction
- Sketch semi-quantitative time profiles of reactions
- Apply extent of reaction concepts to reaction rate considerations
- Use the finite difference method to approximate rates of change of concentrations
- Write rate laws in derivative and integrated from
- Determine the order of reaction by inspection of a rate law expression
- Use the method of initial rates to determine a rate law
- Determine a rate law with the isolation method
- Define the rate-limiting step in a reaction mechanism and describe what is meant by unimolecular, bimolecular, and termolecular reactions
- Apply the Steady State Approximation to reaction mechanism and rate law expressions
- Relate rate constants and equilibrium constants via the Principle of Detailed Balance
- Relate the rate constant and T through Arrhenius's relation $k = A\exp(-E_a/RT)$
- Sketch energy profiles for reactions that indicate the activation energy needed for a reaction
- Compare and contrast collision theory and transition state theory as descriptors of chemical reactions
- Describe selective energy consumption and specific energy disposal
- Use the random walk model to quantify diffusion problems
- Apply activation energy concepts to energy profiles to explain the role of catalysts
- Detail the process of enzyme and surface catalysis

Chapter 15 KEY EQUATIONS:

- 15.3, the equations of Table 15.2, 15.20, 15.27, 15.28, 15.34, 15.35, 15.36, 15.42, 15.51, 15.55, 15.61

Overview

In the introduction to the Chapter 12 solutions, we talked about the fact that "spontaneous" reactions do not have to be fast reactions. We now return to this seemingly strange lack of connection. The two factors that dictate reaction progress are the *thermodynamics* of the reaction and the *kinetics* of the reaction, using rather colloquial terms which, nonetheless, are frequently encountered. Thermodynamics tells us if the reaction is possible. Kinetics tells us how fast the reaction will happen, if it is possible, via the rate constant. Thus, any reaction must be thermodynamically and kinetically favored if it is to proceed quickly. These considerations lead to statements such as: "The explosive water formation reaction is thermodynamically favored, yet it is kinetically slow." It is ironic that thermodynamics is introduced rather early in any chemistry class (as enthalpy), yet, in many ways, (classical) thermodynamics ignores chemistry! Kinetics and its close relative, chemical reaction dynamics, attempt to actually say what the individual molecules, atoms, or electrons are doing during a reaction. Thermodynamics couldn't care less about that. If you are still a little bit uncertain as to what a state variable is, perhaps this difference between thermodynamics and kinetics can help explain it to you. State functions are oblivious to the reaction mechanism and even to the species involved in the reaction, and they only tell you about energetic costs or the current state of a system. Kinetics, however, is much closer to the action. Kinetics attempts to detail how many species collide, if any, during a reaction, and kinetics dares to push electrons around in mechanisms. The concepts of kinetics and chemical reaction dynamics are much less abstract and more chemistry-like than the concepts of thermodynamics. (Remember entropy?) As vital as thermodynamics has been, it is not the science of those who ask questions that start with "How". Do not, however, think that we are advocating a "kinetic view" of a reaction over a "thermodynamic view", analogous to one who subscribes to the Copenhagen interpretation of quantum mechanics. Indeed, this is impossible. *Both* aspects must be understood to fully explain a reaction; *thermodynamics and kinetics are complementary aspects of a reaction.*

Exercises 1, 2, and 3 require that an extent of reaction analysis be performed, and Equation 15.3 relates the reaction rate to the concentrations of the reactants and products as the reaction proceeds. For Exercise 4 plotting $\ln c$ vs t gives a straight line, indicating a first order reaction. With today's calculators you can very quickly plot such a graph, even during an exam, but the finite difference method is more than acceptable. Note also that there is a factor of 2 in front of the lone reactant of Exercise 4 that is not present in Equation 15.1, and you should find that $t_{1/2} = (\ln 2)/2k$ from performing the integration of Equations 5.5 through 5.7 again. This same factor of 2 occurs in Exercise 6. Note that our extent-of-reaction x approach is responsible for the factor of 2 in both Exercises. Additionally, this approach changes the definition of the rate constant. The integrated expressions that we derive in Exercises 4 and 6 in no way contradict Table 15.2. If we were to rewrite the N_2O_5 and HI decompositions for one mole of reactant, this factor of two would be gone, and the integration would proceed *exactly* as in the text examples and the integrated rate laws of Table 15.2 would be found. Obtaining a rate law for a reaction is much more involved than obtaining Q or calculating ΔG, ΔS, and ΔH from the usual "product – reactant" equations. We specifically mention these operations because the coefficients of the balanced equation play such a crucial role in completing them. For rate law considerations, however, no such definite relationship exists. Only if a reaction is found to be elementary, do the exponents of the rate law always equal the stoichiometric coefficients. Further, reactants that appear in the balanced equation are sometimes "missing" from the rate law (OH^- in Example 15.2), and exponents can even be negative ($[O_2]^{-1}$ in Exercises 9 and 17) or fractions (Exercise 10). Only through the information such as we find in Exercises 7 through 14 can one obtain the exponents of a rate law. For example, Exercises 13 and 14 are method of initial rate problems. Additionally, from solving Exercises 6 through 9 one can obtain the units for a first, second, third, and fourth order rate constant. Even if you are not assigned any of these Exercises, you will reap a chemical behoof if you examine closely the units on each k.

Section 15.2 contains differential and integral calculus, and Section 15.3 contains more differentials and equations such as Equation 15.27, indicating why this material is not typically presented early in a chemistry course, as we mentioned above. The Steady State Approximation allows us to analytically solve a set of rate equations by setting the rate of change of the concentrations of the intermediates equal to zero. Even the analytical solutions, however, can be difficult (Exercise 24) or nearly intractable (Exercise 23). Reducing the various rate equations that result from a multi-step mechanism to a single rate equation with the SSA is a way for chemists to determine if the slow step, the rate-limiting step, has been correctly identified. (Identified in that odd sense of "identify" since no reaction mechanism can ever be positively confirmed.) This slow step confirmation is achieved if the rate law that results from the bottleneck step equals the rate law that results from application of the SSA to the rate equations for the entire mechanism. Consider Exercise 20 in which you are asked to consider the reverse of the first step of the proposed mechanism in Example 15.4. This will change the *derived* SSA rate law, but this newly derived rate law will still agree with the *observed* rate law because $k_{-1} \ll k_2[Br_2]$.

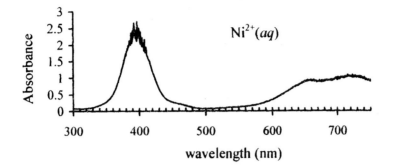

c.   The energy level difference is given in Table 17.1. For Co(II) the spacing is 19,600 cm^{-1}, using $\lambda_{max} = 510$ nm. For Ni(II) the spacing is 13,900 cm^{-1}.

15. The structures below show that the planar geometry generates *cis-* and *trans-* isomers, whereas the tetrahedral geometry exists as only one isomer. *cis-* and *trans-* isomers were discussed in Example 6.4 are discussed further in Chapter 18.

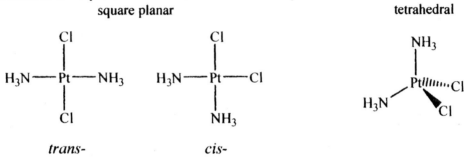

17. a.   The larger K_f is, the more stable is the compound that is formed. Table 17.3 then dictates the order given in the text answer key. Aqua complexes don't appear in Table 17.3 because the table was developed relative to the stability of H_2O complexes.

    b.   Since H_2O will displace Cl^-: $CuCl_4^{2-} + 6H_2O \rightarrow Cu(H_2O)_6Cl_2 + 2Cl^-$.

19. Those compounds that will show a change in magnetism are those that experience a change in the number of unpaired electrons. The magnetism decreases across the spectrochemical series since the series moves from weak to strong field ligands; it moves from high spin states to low spin states. For the sake of completeness we consider all of the possible d configurations for octahedral complexes. There are seven cases for which there will be no change. d^0 has no electrons. d^1, d^2, and d^3 will always have their spins unpaired, regardless of the ligands present. Also, d^8, d^9, and d^{10} can't experience a change in the number of unpaired electrons. The d^4 through d^7 configurations can experience a change. The d^n configurations are: a) d^6 b) d^4 c) d^6 d) d^2 e) d^8 f) d^0 g) d^7 h) d^5 i) d^3 j) d^5. Thus, Co^{3+}, Cr^{2+}, Fe^{2+}, Co^{2+}, Fe^{3+}, and Mn^{2+} will experience a change in magnetism. The number of unpaired spins for the cases of no change are shown in the following table:

d^n	unpaired electrons
d^0, d^{10}	0
d^1, d^9	1
d^2, d^8	2
d^3	3

18

The Chemistry of Carbon

$$\frac{d[ML_6^{n+} \text{---} M'L_6'^{m+}]}{dt} =$$

$$k_{OC}[ML_6^{n+}][M'L_6'^{m+}] - k_{-OC}[ML_6^{n+} \text{---} M'L_6'^{m+}] - k_e[ML_6^{n+} \text{---} M'L_6'^{m+}] \approx 0$$

which, when solved for $[ML_6^{n+} \text{---} M'L_6'^{m+}]$, yields

$$[ML_6^{n+} \text{---} M'L_6'^{m+}] = \frac{k_{OC}[ML_6^{n+}][M'L_6'^{m+}]}{k_{-OC} + k_e}$$

Substitution of this expression into the slow step rate law gives the SSA rate of

$$rate_{k_e} = k_e \left(\frac{k_{OC}[ML_6^{n+}][M'L_6'^{m+}]}{k_{-OC} + k_e} \right)$$

This reduces to Equation 17.10 if $k_e \ll k_{-OC}$, since the denominator becomes simply k_{-OC}. Equation 17.12 takes care of the constants.

33. The f-block elements all have ions with valence electron configurations $(n-2)f^y$. Be sure you have the atomic configuration correct first! See the text answer key, which illustrates why samarium is used in magnets.

35. U has 2 valence electrons $(7s^2)$, and O has 6. If the ion is 2+, there must be twelve valence electrons in a reasonable Lewis structure. See the text answer key. A valence bond description is difficult, but if we assume that the $7s$ and $7p$ orbitals are hybridized, the bonding is the same as we see in CO_2.

d.   From Figure 7.8 O_2: $(\sigma_{2s})^2(\sigma_{2s}*)^2(\sigma_{2p})^2(\pi_{2p})^4(\pi_{2p}*)^2$. Here, there is a tricky problem. The $\pi_{2p}*$ MO is ½ filled. Thus, some stabilization *might* be possible, but with N_2 being the poor back-bonder that it is, O_2 is likely to be an even more inert ligand.

e.   From Figure 7.8 Br_2: $(\sigma_{2s})^2(\sigma_{2s}*)^2(\sigma_{2p})^2(\pi_{2p})^4(\pi_{2p}*)^4$. The $\pi_{2p}*$ orbital is filled, and no back-bonding delocalization into the orbital is possible.

f.   The σ network of this planar compound uses six electrons, leaving two electrons. The two unhybridized p orbitals on C and O combine to give π_p and π_p* MO's. The occupation is $(\pi_p)^2$, indicating that there is an empty π_p* MO. Thus, back-bonding could occur with CH_2O.

g.   From Figure 7.11 NO: $(\sigma_s)^2(\sigma_s*)^2(\pi_p)^4(\sigma_p)^2(\pi_p*)^1$. This nearly vacant π_p* orbital allows for some back-bonding. However, it is not as successful as CO. Proof being that $Cr(NO)_4$ is the only metal complex that forms exclusively with NO ligands. Thus, it is like N_2 and always (except for the case just mentioned) bonds in the presence of other ligands.

h.   From Figure 7.11 NO^+: $(\sigma_s)^2(\sigma_s*)^2(\pi_p)^4(\sigma_p)^2$. This molecule is isoelectronic with CO, and it, therefore, has an empty π_p* orbital. We predict that it will back-bond. The sketch of CN⁻'s back-bonding interaction is shown in Figure 17.14. The case of CO is nearly identical since the two are isoelectronic. Note, carefully, the orbital amplitude is larger on the C atom in the π_p* orbital, and the ligand bonds to the metal at the C atom. (See also Figure 7.12.)

29. a.   If asked to make the mechanism analogous to Equation 17.7, let $n = 6$ for the reverse reaction, as well. The reaction we consider, therefore, is $ML_5X + L \rightarrow ML_6 + X$. The analogous mechanism is

$$ML_5X + L \underset{k_{-OC}}{\overset{k_{OC}}{\rightleftharpoons}} [ML_5X\text{---}L]$$

$$[ML_5X\text{---}L] \xrightarrow{k_L} [ML_5\text{---}L] + X$$

$$[ML_5\text{---}L] \xrightarrow{k} ML_6$$

b.   For the forward reaction, in which an M–L bond is broken, $E_a \approx D(M–L)$. For the reverse reaction an M–X bond is broken. So, $E_a' \approx D(M–X)$ is expected. The forward and reverse E_a's are related to the heat of reaction by Equation 15.36. That is, $E_a' = E_a - \Delta H°$.

c.   Since $D(M–X) >> D(M–L)$, $E_a' >> E_a$. Therefore, the rate of the reverse reaction will be much less than the rate of the forward reaction.

31. The slow step is the second step for which the rate law is $\text{rate}_{k_e} = k_e[ML_6^{n+} \text{ - - - } M'L_6'^{m+}]$.

Thus, we need to find an expression for this intermediate from the reactants or products. Being sure to include the reverse of the equilibrium in the first step, the rate of change of concentration of the associated outer-sphere intermediate is

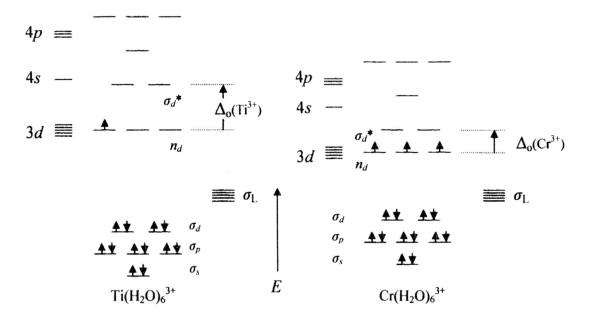

25. a. The detailed solution to this type of problem is shown for Exercise 25, Chapter 14. Please refer to it. See the text answer key.

b. While a smaller radius is *very* often cited as a reason for less overlap (e.g., the $1s$ orbitals not overlapping significantly in the MO description of the $n = 2$ diatomic molecules), orbital radius is not the only measure or predictor of what leads to successful overlap. Smaller orbitals are more dense, allowing for more effective overlap. There exists, then, a competition between how effective the overlap is between larger, but more diffuse, orbitals and smaller, but more dense, orbitals. This idea is clearly at the center of why C–C bonds are much stronger than Si–Si bonds. See the discussion of silicon at the end of Section 18.1. For the case of these iron complexes the denser Fe^{3+} orbital has more successful overlap, that is, stronger covalent bonding, with the water ligands, a fact reflected in the larger Δ_o of the previous Exercise.

27. As the text points out, ligands with filled π symmetry orbitals destabilize the t_{2g} level, decrease Δ_o, and destabilize the complex. Ligands with an empty π^* symmetry orbital, however, will stabilize the t_{2g} level, increase Δ_o, and stabilize the complex as shown in Figure 17.14. This stabilization is due to "back-bonding." (As usual, though, the orbitals must be of similar energy in order to interact sufficiently. If a ligand has an empty π^* symmetry orbital that is very different in energy than the t_{2g} orbitals, then no significant interaction can occur.) Thus, an examination of the MO occupations is in order.

a. H_2 has no π MO's. See Figure 7.3. No back-bonding is possible.

b. From Figure 7.11 CO: $(\sigma_s)^2(\sigma_s^*)^2(\pi_p)^4(\sigma_p)^2$. Thus, it has an empty π_p^* MO. CO is, in fact, the most important example of a ligand that exhibits back-bonding.

c. From Figure 7.8 N_2: $(\sigma_{2s})^2(\sigma_{2s}^*)^2 (\pi_{2p})^4(\sigma_{2p})^2$. Again, there is an empty π_{2p}^*. However, N_2 is nowhere near as active as a ligand as CO. In fact, it only bonds in the presence of other ligands. A major factor in this difference is the small difference in the MO energies of CO and N_2, as mentioned parenthetically above.

For the cases where there is a change in the magnetism:

d^n	unpaired electrons in high spin case	unpaired electrons in low spin case
d^4	4	2
d^5	5	1
d^6	4	0
d^7	3	1

Convince yourself that these are the correct numbers of electrons by filling in the t_{2g} and e_g levels for each case. Note that the number of unpaired electrons decreases in the presence of strong field ligands. With fewer unpaired spins the magnetic properties of the compound decrease.

21. See Figure 17.13. The energy level diagrams for this problem are very similar. There are still 12 electrons from the six ligands that occupy the lowest energy MO's. The are, however, eight electrons from $Ni^{2+}(d^8)$ that now occupy the n_d and σ_d^* levels. From Figure 17.9 $\Delta_o(NiCl_6^{4-}) \approx 12,800$ cm^{-1} and $\Delta_o(Ni(NH_3)_6^{2+}) \approx 16,900$ cm^{-1}. Thus, the enhanced stability of the ammine complex should be reflected in your diagram by its larger Δ_o, which exists because the bonding orbitals are lower in energy (and the anti-bonding orbitals are higher in energy) in the ammine complex due to better orbital overlap and stronger covalent bonding. [The keen observer might think that there is some sleight of hand going on here. Perhaps, legerdemain. We say that the bonding MO's decrease in energy, and this leads to greater stability. But, at the same time the energy of the anti-bonding orbitals increases. Shouldn't these effects cancel? They would if the anti-bonding orbitals were filled. This is only the case for d^{10} compounds. And, indeed, Zn^{2+} does not form stable octahedral compounds.]

23. The energy level diagrams are below. The lines connecting the AO's and the MO's were omitted for clarity. The difference in color is due to a larger Δ_o in the Ti^{3+} complex than in the Cr^{3+} complex. From Table 17.1 this larger energy difference results in absorption at 560 nm as opposed to 610 nm. Based on the level of presentation in this chapter, one could reasonably pick either compound as the more stable one. As discussed in Exercise 21, a larger Δ_o indicates greater covalent bonding and, consequently, a more stable complex. Thus, the Ti^{3+} complex should be more stable. However, Cr^{3+} has a d^3 configuration, while Ti^{3+} is d^1. In the crystal field model subsection of Section 17.4, we noted that d^3 configurations are especially stable. Thus, Cr^{3+} should be more stable. There is no contradiction in the presentation of Chapter 17. Both the Δ_o spacing and d^3 stability facts are true. The problem lies in the simplicity of the models that have been used in this introductory chapter. (Kinetic studies of ligand substitution, as described in Section 17.5, have shown that the rate of H_2O substitution is very slow for $Cr(H_2O)_6^{3+}$ because the $Cr^{3+}-OH_2$ bonds are very stable.)

Note in the figure that the AO levels fall for Cr^{3+} due to its larger Z, as in Figure 17.6. Thus, we have the unusual situation where the resonance is better (i.e., the energy is more closely matched [see Section 7.2 and Mulliken's rule II]) between the metal ion AO's and the ligand orbitals in $Cr(H_2O)_6^{3+}$ than in $Ti(H_2O)_6^{3+}$, yet the bonding is less covalent. Knowing the great stability of $Cr(H_2O)_6^{3+}$, factors other than covalent bonding must play a role in determining the final energetics of these two compounds.

Industrial chemical companies would rank Equation 15.35 as one of the most important equations in *University Chemistry*. The entire "game" in industry is find a set of optimal conditions that leads to high yields quickly. Exercises 12 through 15 in Chapter 12 explored the thermodynamic aspect of improving industrial reactions, and Equation 15.35 is the central theoretical equation to describe the kinetic aspect of these, or any other, reactions. This equation is easily shown to be equivalent to Equation 12.15 in Example 15.5, and in this form it is used to solve Exercises 25 and 27. Exercise 26 is first a detailed balance Exercise which must be solved for $k' = k/K$ with K = 131 atm[(0.08206 L atm/K mol)803 K]$^{-1}$. A' follows from Equation 15.35. For Exercise 28(a) a graph of $\ln k$ vs $1/T$, which is the graph of the line $\ln k = \ln A - E_a/RT$, gives a slope equal to $- E_a/R$ and a y-intercept equal to $\ln A$. For Exercise 28(b) note that $\Delta H°_{298}(I_2(g)) \neq 0$, as the reference form is solid at 298 K.

The topics of collision theory, transition state theory, and molecular reaction dynamics are beyond any typical general chemistry course. This doesn't mean that you are not capable of learning this material, but it will not be "a walk in the park". The typical stumbling block in chemistry and physics is the math, but if you take a glimpse at the equations in Section 15.4, you will find that these equations, or their analogous forms, have been encountered in previous chapters. Transition state theory is more complex that collision theory, but it is more pleasing to the analytical and theoretical crowd in that it combines thermodynamics with kinetics. Note that for diffusion-controlled reactions (in solution), the rate at which the reaction proceeds depends on how fast the reactants can get to each other. This "getting to each other" is called diffusion, and the velocity at which this occurs is given by Equation 15.56. It is worth pointing out that the flow of electrons through wires is analogous to the diffusion of reacting ions through solution. The electrons have an effective speed of 10^6 m/s, but the collisions they make with the nuclei, ignoring any electron interactions, gives them a short mean free path and very small *drift speed* which is analogous to diffusion velocity in solutions. For Exercise 38, solve Equation 15.55 for Δt, and for Exercise 40 use $d_{AB} = 2(1.73$ Å$)$ because r_{vdW} is a *radius*.

Catalysis is a vital process in the truest meaning of the word vital. We wouldn't exist without the catalysts in our bodies called enzymes. See Figures 15.18 and 18.21. Whether the catalyst is in a biological system or not (Figure 15.17), however, a catalyst lowers the activation energy for a reaction by allowing it to proceed *via a different pathway*. The inset of Figure 18.21 shows this explicitly. For Exercise 46 see Figure 15.19.

1. a. Given that there is a stoichiometric mixture, there are 5/2 as many moles of $H_2C_2O_4$ as moles of MnO_4^-. Therefore,

$$2MnO_4^- \quad + \quad 5H_2C_2O_4 \quad + \quad 6H^+ \quad \rightarrow \quad 2Mn^{2+} \quad + \quad 10CO_2 \quad + \quad 8H_2O$$

initially	c_0	$(5/2)c_0$	excess	0	0	-
at time t	$c_0 - 2x$	$(5/2)c_0 - 5x$	excess	$2x$	$10x$	-

Now, the reaction rate is equal to dx/dt. From Equation 15.3

$$\text{reaction rate} = \frac{dx}{dt} = -\frac{1}{2}\frac{d[MnO_4^-]}{dt} = -\frac{1}{5}\frac{d[H_2C_2O_4]}{dt} = \frac{1}{2}\frac{d[Mn^{2+}]}{dt}$$

In words, Mn^{2+} is formed at the same rate that MnO_4^- disappears, and $H_2C_2O_4$ disappears at a rate 5/2 times that of MnO_4^-.

b. A time profile *sketch* is below. Due to the different rates of disappearance, the slope of the $H_2C_2O_4$ curve is greater than that of MnO_4^-. $[Mn^{2+}]_{eq}$ approaches 2 M asymptotically.

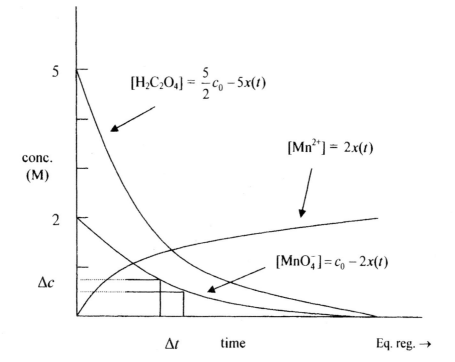

$$[H_2C_2O_4] = \frac{5}{2}c_0 - 5x(t)$$

$$[Mn^{2+}] = 2x(t)$$

$$[MnO_4^-] = c_0 - 2x(t)$$

conc. (M)

Δc

Δt time   Eq. reg. →

3. a. Letting $[O_3]_0 = c_0$, we get the following extent of reaction analysis

$$2O_3 \quad \rightarrow \quad 3O_2$$

initially	c_0	0
at time t	$c_0 - 2x$	$3x$

From Equation 15.3

$$\text{reaction rate} = \frac{dx}{dt} = -\frac{1}{2}\frac{d[O_3]}{dt} = \frac{1}{3}\frac{d[O_2]}{dt}$$

b. The sketch is below. Note that $[O_3]_{eq}$ goes to zero, and $[O_2]_{eq}$ goes to $1.5c_0$.

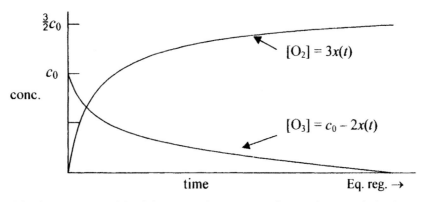

c. For ideal gases, $P = cRT$. Of course, the pressure for a mixture of ideal gases depends on the number of moles present (not their identity), which at time t is given by $(c_0 - 2x) + 3x = c_0 + x$, from the above extent of reaction analysis. As in Example 15.1, $P = (c_0 + x)RT$. Remember that the value in parentheses is a concentration, and the units, therefore, work out.

5. a. As in Exercise 4 and Figure 15.2, plotting the following points ($\ln c$ vs. t) gives a straight line: (Let's forget the infinity point.)

time	$\ln c$
0	1.89
360	1.77
720	1.66
1080	1.55

Thus, the reaction is first order (see Table 15.2), and linear regression gives the line: $\ln c = -3.14 \times 10^{-4} t + 1.89$. Thus, reaction rate $= -d[D]/dt = k[D]$. Since the reactant stoichiometric coefficient is 1, the derivation of k and $t_{1/2}$ runs exactly as shown in the text for C_2H_5Cl. (This is unlike Exercise 4.) Therefore, the slope of the line we just calculated is equal to $-k$, and $k = 3.14 \times 10^{-4}$/s. Equation 15.8 gives $t_{1/2} = \ln 2/(3.14 \times 10^{-4}$/s$) = 2207$ s ≈ 2200 s.

b. The time profiles *sketches* are below. From the given data, there is almost complete conversion to products. Thus, from the fact that $[D]_{eq} = 0.01$ M and the stoichiometry of the balanced equation, $[E]_{eq} = 6.61$ M and $[A]_{eq} = 13.22$ M. By definition the concentration of D at $t_{1/2}$, $[D]_{1/2}$, is just half of its initial value, giving $\frac{1}{2}[D]_0 = [D]_{1/2}$. From the stoichiometry of the given reaction, the following relation must also be true at $t_{1/2}$: $\frac{1}{2}[D]_0 = [D]_{1/2} = [E]_{1/2} = \frac{1}{2}[A]_{1/2}$. These $t_{1/2}$ values (3.31 M and 6.62 M) are shown on the sketch.

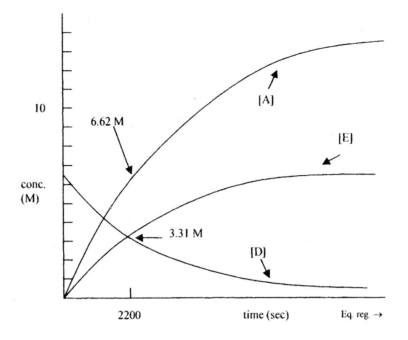

10

6.62 M

conc.
(M)

3.31 M

[A]

[E]

[D]

2200   time (sec)   Eq. reg. →

c. c is the concentration of D at time t. At 90 % decomposition $c = 0.662$ M. Now, since the stoichiometric coefficient for D, the lone reactant, is the same as for C_2H_5Cl in the example solved in the text, Equation 15.7 gives $0.662 = 6.62\exp(-kt_{0.90})$ with $k = 3.14 \times 10^{-4}$/s. Taking the ln of both sides yields $t_{0.90} = 7333$ sec ≈ 7330 s.

7.   The order with respect to each reactant is equal to the power to which the reactant concentration is raised in the rate law. The overall order is the sum of the exponents. Thus, we get second, first, and third order. Now, the reaction rate has the unit mol/L s. (See, for example, Equation 15.2, where rate $= d$(conc.)$/dt$.) The unit obtained from $[NO]^2[H_2]$ is mol^3/L^3. To get the right side to equal the left side, k must have the unit of L^2/mol^2 s. Because the reaction is second order with respect to [NO], doubling [NO] quadruples the rate, a factor of four. Because the reaction is first order with respect to $[H_2]$, halving $[H_2]$ halves the rate, a factor of ½. Tripling both concentrations increases the rate by a factor of $9 \times 3 = 27$.

9.   This rate expression shows that not all orders are positive. Doubling $[O_3]_0$ will quadruple the rate if the reaction is second order with respect to O_3. For O_2, doubling $[O_2]_0$ can *decrease* the rate by a factor of two only if the reaction is -1 order with respect to O_2. The rate law is reaction rate $= k[O_3]^2[O_2]^{-1}$, and the overall reaction is first order. The product of the concentrations will have the unit mol/L. If the reaction rate is to have the unit of mol/L s, the rate constant must have the unit of s^{-1}.

11.   This problem serves as a precursor to the many mechanism and steady state approximation exercises that include back reactions.   An extent of reaction analysis:
$$C_2H_5Cl \;\rightarrow\; C_2H_4 \;+\; HCl$$

initially	c_0	0	0
at time t	$c_0 - x$	x	x

The rate of the forward reaction is given by ($c = [C_2H_5Cl]$)
$$\text{reaction rate}_{for} = -\frac{dc}{dt} = kc = k(c_0 - x)$$

In terms of x for the reverse reaction

$$\text{reaction rate}_{rev} = \frac{dc}{dt} = k'(x)(x) = k'x^2$$

Now, since the net rate is equal to the forward rate minus the reverse rate, $dx/dt = k(c_0 - x) - k'x^2$. Lastly, the time profile would be modified in that the concentration of the products would not reach c_0, nor would the reagent concentration go to zero. This is because the equilibrium does not lie as far to the right, and the final concentrations of C_2H_4 and HCl can't be as high as they were at the higher temperature.

13. This is the method of initial rates. We assume that the rate law is of the form

$$\text{reaction rate} = -\frac{d[Br_2]}{dt} = k[AH]^p[Br_2]^q[H^+]^r$$

The ratio of the rates for runs I and II (only $[AH]_0$ changes) determines p:

$$\frac{\text{rate I}}{\text{rate II}} = \left(\frac{[AH]_I}{[AH]_{II}}\right)^p \quad \Rightarrow \quad \frac{5.11 \times 10^{-5} \text{mol/L s}}{2.64 \times 10^{-5} \text{mol/L s}} = \left(\frac{4.0 \text{ M}}{2.0 \text{ M}}\right)^p$$

(The above rates are calculated from the time and initial concentration data for Br_2 since it is the color of Br_2 that is monitored as the reaction proceeds. For example, rate I = 0.0024 M/47 s = 5.11×10^{-5} mol/L s.) Thus, $p = \ln(1.94)/\ln(2.0) = 0.953 \approx 1$. To calculate q, compare runs I and IV, which has a rate of 4.95×10^{-5} mol/L s. For these two runs only $[Br_2]_0$ has changed. Note, however, the ratio of the rates for run I and run IV gives $1.03 \approx 1$; the rate does not change when the concentration of Br_2 is changed. As for OH⁻ in Example 15.2, this indicates the reaction is zero order with respect to Br_2. Explicitly calculating, $\ln(1)/\ln(\frac{1}{2}) = q = 0$. A comparison of runs I and III, which has a rate of 1.04×10^{-4} mol/L s, determines r. Solving for r gives $r = \ln(0.49)/\ln(0.5) \approx 1$. Thus, reaction rate = $k[AH][H^+]$. Finally, using data for run I

$$k = \frac{5.11 \times 10^{-5} \text{ mol/L s}}{(4.0 \text{ M})(1.0 \text{ M})} = \frac{1.28 \times 10^{-5} \text{ L}}{\text{mol s}}$$

For run II, $k = 1.32 \times 10^{-5}$ L/mol s, and run III gives $k = 1.30 \times 10^{-5}$ L/mol s. The average of these three k values is $k_{avg} = 1.30 \times 10^{-5}$ L/mol s.

15. a. An elementary reaction occurs as it is written – in one step. Step 1, then, occurs from one molecule breaking up. Steps 2 and 3 occur from two molecules actually colliding. Thus, step 1 is unimolecular, and steps 2 and 3 are bimolecular. The rate laws are given in the text key. Note that for elementary reactions the order of the reaction with respect to each reactant is equal to the stoichiometric coefficient. The three steps sum to the overall reaction since the $(CH_3)_3C^+$, H_2O, and H^+ all cancel.

    b. The bottleneck principle says that the observed rate law for the overall reaction will depend only on the slow step. Step one is the slow step of this mechanism, and the rate law found for step one agrees with Equation 15.22. Therefore, it agrees with observation. This also means that a mechanism is only unique for the rate-limiting step.

17. Each step in a mechanism is elementary, indicating that step 1 is unimolecular since it has one reactant, O_3. The reverse of step 1 is bimolecular since it has two reactants, O_2 and O. Step two has two reactants, and it is bimolecular. We verify that these steps add up to the overall reaction without using the reverse of step one. Between the two steps the O atoms cancel, yielding the overall reaction. The bottleneck principle says that the observed rate law for the overall reaction depends only on the slow step. Here, the slow step is step two, for which rate$_2 = k_2[O_3][O]$. Next, as in Exercise 16, note that the first step is at equilibrium, meaning $K_1 = [O_2][O]/[O_3]$. This rearranges to $[O] = K_1[O_3]/[O_2]$. Substitution of this into

the $rate_2$ expression gives $k_2K_1[O_3]^2[O_2]^{-1}$. This equals the observed rate law for the overall reaction if $k_2K_1 = k$. Lastly, since the first step is at equilibrium, LeChâtelier's principle can be used. Increasing the O_2 concentration drives the step 1 equilibrium to the left, decreasing [O]. Since $rate_2$ depends on [O], this will reduce the rate of the rate-limiting step.

19. To verify, simply cancel all of the intermediates that appear on both sides of the equations. The bottleneck principle says that the overall reaction rate law is determined from the slow step. Step 3 gives $rate_3 = k_3[H_2I][I]$. Now, unlike Exercises 16, 17, and 18, *two* intermediates are present in the rate-limiting step. Thus, two expressions, one for $[H_2I]$ and a second for $[I]$, are needed. The expression for $[H_2I]$ comes from the equilibrium established in step 2. From this $K_2 = [H_2I]/[I][H_2]$, or $[H_2I] = K_2[I][H_2]$. One must be careful not to find the expression for $[I]$ from this same equilibrium. Doing so would give a product of one for $[H_2I][I]$. The expression for $[I]$ comes from the first equilibrium: $K_1 = [I]^2/[I_2]$, or $[I]^2 = K_1[I_2]$. Substituting our expression for $[H_2I]$ into $rate_3$ gives $rate_3 = k_3K_2[I][H_2][I] = k_3K_2[I]^2[H_2]$. Next, substitute $[I]^2$ into $rate_3$ to get $rate_3 = k_3K_2K_1[I_2][H_2]$. This is consistent with the observed rate law if $k_3K_1K_2 = k$.

21. The reactions to consider and their rate laws are:

$$1. \quad O_3 \rightarrow O_2 + O \qquad\qquad rate_1 = k_1[O_3]$$
$$-1. \quad O_2 + O \rightarrow O_3 \qquad\qquad rate_{-1} = k_{-1}[O_2][O]$$
$$2. \quad O_3 + O \rightarrow O_2 + O_2 \quad rate_2 = k_2[O_3][O]$$

Now, as was done with E in Equation 15.27, O will be expressed in terms of O_3 and O_2 and substituted it into the $rate_2$ expression, since this is the slow step in the mechanism. Since O is present in all three steps, there are three terms in our expression for $d[O]/dt$. The step where O is formed gives a (+) term, and the steps where O is consumed give a (−) term, giving $d[O]/dt = k_1[O_3] - k_{-1}[O_2][O] - k_2[O_3][O]$. This is approximately zero in the SSA. Setting this expression equal to zero and solving gives

$$[O] = \frac{k_1[O_3]}{k_{-1}[O_2] + k_2[O_3]}$$

This expression can now be substituted into $rate_2$ to give the overall reaction rate:

$$\text{reaction rate} = k_2[O_3]\left(\frac{k_1[O_3]}{k_{-1}[O_2]+k_2[O_3]}\right) = \frac{k_2k_1[O_3]^2}{k_{-1}[O_2]+k_2[O_3]}$$

Now, if $k_{-1}[O_2] \gg k_2[O_3]$, the SSA and observed laws are equal, provided that $k_2k_1/k_{-1} = k$. This inequality is justified for two reasons. First, the step 1 equilibrium lies far to the left due to the bond energy of O_3 that inhibits dissociation. Thus, k_{-1} is large, and recombination is fast. Second, step 2 is slow, implying that k_2 is small. Lastly, Exercises 9 and 17 refer to laboratory (not atmospheric) conditions. Evidently, differences in $[O_2]$ and $[O_3]$, however large or small, are not great enough to invalidate our inequality or to obscure the overriding importance of the disparate k values.

23. The five chemical equations that we need are in Exercise 19. We are attempting to derive an expression for $d[HI]/dt$ in terms of I_2 and H_2 by finding expressions for the intermediates of the reaction, I and H_2I. The rate laws for each step are

$$\text{rate}_1 = -\frac{d[I_2]}{dt} = \frac{1}{2}\frac{d[I]}{dt} = k_1[I_2]; \qquad \text{rate}_{-1} = -\frac{1}{2}\frac{d[I]}{dt} = \frac{d[I_2]}{dt} = k_{-1}[I]^2;$$

$$\text{rate}_2 = -\frac{d[I]}{dt} = \frac{d[H_2I]}{dt} = k_2[I][H_2]; \qquad \text{rate}_{-2} = -\frac{d[H_2I]}{dt} = \frac{d[I]}{dt} = k_{-2}[H_2I];$$

$$\text{rate}_3 = -\frac{d[H_2I]}{dt} = -\frac{d[I]}{dt} = k_3[H_2I][I]$$

[I] appears in all five rate equations, and [H$_2$I] appears in three of them. Solving all of the above rate expressions for $d[I]/dt$ and adding them leads to the expression in the text key. Likewise for $d[H_2I]/dt$. Once found, the expressions for [I] and [H$_2$I] are substituted into rate$_3$, the rate limiting step. As this problem states, however, the solution for this problem is formidable. Unfortunately, [I] occurs as $[I]^2$, and the analytical solution is nearly interminable and not worth our time at the present. In the next Exercise [Br] occurs as $[Br]^2$, too. However, a key cancellation makes an analytical solution possible without (too much) anguish. Let's focus on that success and not the analytical difficulty that we have here!

25. a. Example 15.5 contains a different form of Equation 15.35, identical to the integrated form of the van't Hoff equation. To find E_a, solve

$$\ln\left(\frac{0.0112\,s^{-1}}{4.20\times10^{-5}\,s^{-1}}\right) = \frac{-E_a}{0.0019872\,\text{kcal/K mol}}\left(\frac{1}{803\,K} - \frac{1}{703\,K}\right)$$

So, $E_a = 62.7$ kcal/mol ≈ 262 kJ/mol. Using the k/T pair for 803 K in Equation 15.35 gives

$$A = \frac{0.0112\,s^{-1}}{\exp\left(\dfrac{-62.7\,\text{kcal/mol}}{(0.0019872\,\text{kcal/K mol})803\,K}\right)} = \frac{1.30\times10^{15}\,\text{collisions}}{\text{sec}}$$

   b. To solve this, use the same equation that was used in part (a), where the unknown is now k_{298}. This gives

$$\ln\left(\frac{4.20\times10^{-5}\,s^{-1}}{k_{298}}\right) = \frac{-62.7\,\text{kcal/mol}}{0.0019872\,\text{kcal/K mol}}\left(\frac{1}{703\,K} - \frac{1}{298\,K}\right)$$

Taking the exponential of both sides leads to $k_{298} = 1.36 \times 10^{-31}$/s, an extremely slow rate which accounts for the lack of experimental evidence at 298 K. This can be understood in terms of the "Boltzmann factor," $\exp(-E_a/RT)$. As mentioned before Equation 15.41, this is the fraction of molecules with enough energy to react, the molecules with energy equal to or exceeding E_a. (Remember that it is a ratio, as shown in Equations 10.34 and 10.35.) For $E_a = 62.7$ kcal/mol, this gives a fraction of 1.1×10^{-46}. Thus, in a one mole sample, where there are 6.0221×10^{23} molecules, not one molecule will have enough energy to decompose. For example, if *one* molecule had enough energy to decompose, the fraction would be $1/(6.0221 \times 10^{23}) = 1.7 \times 10^{-24}$. The Boltzmann factor here is 22 orders of magnitude smaller than that.

   c. Below Equation 15.1 $\Delta H°$ is given as 17.24 kcal. Thus, this reaction is endothermic, and E_a' will be less than E_a via Equation 15.36. From this equation $E_a' = 62.7 - 17.24 = 45.5$ kcal/mol. The energy profile *sketch* (not to scale) is shown below, where it is easy to see that $\Delta H°$ is the difference between the E_a and E_a'. (The small difference between E and H is ignored.)

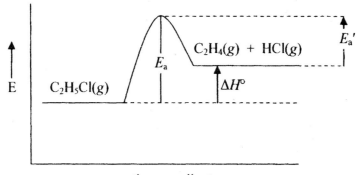

reaction coordinate

27. This Exercise is easily solved using the modified form of Equation 13-35 that is found in Example 15.5. Since the rate doubles, $k_2 = 2k_1$. Thus,

$$\ln\left(\frac{2k_1}{k_1}\right) = \frac{-E_a}{0.0019872\,\frac{kcal}{K\,mol}}\left(\frac{1}{308\ K} - \frac{1}{298\ K}\right) \Rightarrow E_a = \frac{12.6\,kcal}{mol}$$

29. The equations of interest are $O + CO_2 \rightarrow O_2 + CO$ and its reverse. Now, ΔH°_{298} from Appendix C for $O + CO_2 \rightarrow O_2 + CO$ is 8.082 kcal, so this reaction is endothermic with $E_a = 54.2$ kcal/mol (from Table 15.4). For endothermic reactions $E_a' < E_a$. From Table 15.4 E_a' for this reaction is 51.0 kcal/mol, and this means that the activation energies are consistent with the heat of reaction that we obtained from Appendix C. Note that these activation energies predict a heat of reaction, ignoring the small difference between E and H as in Equation 15.36, of $54.2 - 51.0 = 3.2$ kcal/mol.

The Principle of Detailed Balance can be used to determine K by first calculating k for $O + CO_2 \rightarrow O_2 + CO$ and k' for its reverse. Equation 15.35 gives

$k = (1.91 \times 10^{10}$ L/mol s)exp[$(-54.2$ kcal/mol)$/R(298.15$ K)], and

$k' = (3.47 \times 10^9$ L/mol s)exp[$(-51.0$ kcal/mol)$/R(298.15$ K)].

We found A from $\log_{10}^{-1}$ of the $\log_{10}A$ value listed in Table 15.4. These give $k = 3.57 \times 10^{-30}$ L/mol s and $k' = 1.44 \times 10^{-28}$ L/mol s. So, if $K = k/k'$, then $K = 0.025$. From Equation 12.8, $\Delta G^\circ_{298} = 2.19$ kcal/mol.

Now, the thermodynamic value for $O + CO_2 \rightarrow O_2 + CO$, obtained from Appendix C, is $\Delta G^\circ_{298} = 6.085$ kcal. Equation 12.8 gives $K = 3.5 \times 10^{-5}$. Finally, the thermodynamic data are for $T = 298.15$ K. However, this reaction does not occur to any measurable extent at this temperature. Thus, we actually extrapolated the kinetic data to 298.15 K. Considering this, the agreement is not too bad.

31. In Exercise 28 it was determined that $A = 1.40 \times 10^{11}$ L/mol s. The engineering formula from Exercise 30 (see the text answer key) makes this problem simple, keeping in mind that the end of section 14.4 tells us that the van der Waals radius is half the diameter of the molecule. Thus, $r_{vdW}(HI) = 2.10$ Å, and the diameter is 4.20 Å. Since $d_{AB} = (d_A + d_B)/2$,

$$A_{ct} = 4.5713 \times 10^{-12}\sqrt{\frac{680\ K}{63.957\,amu}}\left(\frac{4.20\ \text{Å} + 4.20\ \text{Å}}{2}\right)^2 \div 2 = \frac{1.32 \times 10^{-10}\ cm^3}{molecule\,s}$$

which can also be expressed as $A_{ct} = 7.92 \times 10^{10}$ L/mol s. The steric factor, p, equals A/A_{ct}. Thus, $p = (1.40 \times 10^{11}$ L/mol s)$/7.92 \times 10^{10}$ L/mol s $= 1.77$.

Now, a steric factor of exactly one means no effect at all, indicating that the experimentally determined A matches the collision theory prediction's A_{ct}. We can approximate that any p on the order of one will describe a situation where there is little steric effect. This range is around $0.5 \rightarrow 2$. Ignoring London dispersion and dipole-dipole attractions, the tiny steric effect present in this collision is explained by the fact that the classical rotational period of the H atom around the molecule's center of mass is short compared to the time for the collision to occur. Therefore, the H atom will always have time to end up in the right place; it is not sterically hindered.

33. As discussed before Example 15.6, the right hand side of Equation 15.53 must be multiplied by $(RT/P^\circ)^{-\Delta n^\ddagger}$ to make the entropy of activation correspond to a standard state of 1 atm ($=$ 101,325 Pa.) Now, $\Delta n^\ddagger = -1$. Substituting this and $(RT/P^\circ)^{-\Delta n^\ddagger}$ into Equation 15.35 leads to (*in SI units*)

$$A = \frac{(1.3807 \times 10^{-23}\ \text{J/K})T}{6.6261 \times 10^{-34}\ \text{J s}} \exp\left(\frac{\Delta S^{\circ\ddagger}}{R}\right) \exp(2)\left(\frac{(8.3145\ \text{J/K mol})T}{101,325\ \text{Pa}}\right)$$

$$= \frac{1.2634 \times 10^7\ \text{m}^3}{\text{K}^2\ \text{mol s}}\,T^2 \exp\left(\frac{\Delta S^{\circ\ddagger}}{R}\right) \times \left(\frac{100\ \text{cm}}{\text{m}}\right)^3\left(\frac{1\ \text{mol}}{6.0221 \times 10^{23}\ \text{molecules}}\right)$$

$$= \frac{2.0980 \times 10^{-11}\ \text{cm}^3}{\text{K}^2\ \text{molecules s}}\,T^2 \exp\left(\frac{\Delta S^{\circ\ddagger}}{R}\right) = \frac{1.2634 \times 10^{10}\ \text{L}}{\text{K}^2\ \text{mol s}}\,T^2 \exp\left(\frac{\Delta S^{\circ\ddagger}}{R}\right)$$

You should convince yourself that the unit after the first step is, indeed, m³/K² mol s. The authors have! In both formulas T must be in K. A_{298} is determined from substituting 298 K in for T. This gives

$$A_{298} = \frac{1.8631 \times 10^{-6}\ \text{cm}^3}{\text{molecules s}} \exp\left(\frac{\Delta S^{\circ\ddagger}}{R}\right) = \frac{1.1219 \times 10^{15}\ \text{L}}{\text{mol s}} \exp\left(\frac{\Delta S^{\circ\ddagger}}{R}\right)$$

Now, the values in Table 15.4 are for 298 K, so the above formula gives

$$A_{298} = \log^{-1}(6.09) = \frac{1.1219 \times 10^{15}\ \text{L}}{\text{mol s}} \exp\left(\frac{\Delta S^{\circ\ddagger}}{R}\right)$$

$$\frac{1.230 \times 10^6\ \text{L}}{\text{mol s}} = \frac{1.1219 \times 10^{15}\ \text{L}}{\text{mol s}} \exp\left(\frac{\Delta S^{\circ\ddagger}}{R}\right) \Rightarrow 1.0966 \times 10^{-9} = \exp\left(\frac{\Delta S^{\circ\ddagger}}{R}\right)$$

Thus, $\Delta S^{\circ\ddagger} = -41.0$ cal/K mol. Note that when taking the inverse log of 6.09 L/mol s, the units are not inverted. The log scale converts the values of A to a log scale. It can't change the unit.

Appendix C gives $\Delta S^\circ = -28.81$ cal/K mol. Now, unlike Example 15.6, the entropy change for the formation of the complex is greater than the entropy change for the whole reaction. Thus, the transition state complex has "more order" than the final product, ethane. This is probably best explained by the existence of a square C_2H_2 ring that is formed as the H_2 adds to the double bond in the ethene.

35. The n's in the formula refer to the number of vibrational degrees of freedom in the colliding species, be they molecules or atoms. Recall from Chapter 8 that there are always $3N$ degrees of freedom (d.f.'s) in a molecule (or atom).

   a. In the linear transition state there are three atoms and nine total d.f.'s. As always for a linear entity, there are three translational d.f.'s and two rotational d.f.'s. This leaves

four d.f.'s in vibration. The single atom, of course, has only three d.f.'s – all translational. The diatomic molecule is necessarily linear and has three translational d.f.'s and two rotational d.f.'s, leaving only one vibrational d.f. Thus, $n_{linear} = 4 - 0 - 1 - 1 = 2$, and $p_{linear} = 0.04$.

The bent transition state has three translational d.f.'s and three rotational d.f.'s. This leaves three d.f.'s in vibration. Everything else is the same, and $n_{bent} = 3 - 0 - 1 - 1 = 1$. Thus, $p_{bent} = 0.2$. Now, the first nine reactions listed in Table 15.4 are of this type. In this rough estimate only the $O + H_2$ and $D + H_2$ reactions are closer to 0.04 than 0.2. Thus, they might have linear transition states. The other seven are predicted to have bent transition states.

b. $\underline{OH + H_2}$: In the linear transition state there are four atoms and 12 total d.f.'s. As always for a linear entity, there are three translational d.f.'s and two rotational d.f.'s. This leaves seven d.f.'s in vibration. The diatomic molecules are necessarily linear and each has three translational d.f.'s and two rotational d.f.'s, leaving only one vibrational d.f. Thus, $n_{linear} = 7 - 1 - 1 - 1 = 4$, and $p_{linear} = 0.0016 \approx 0.002$.

The bent transition state has three translational d.f.'s and three rotational d.f.'s. This leaves six d.f.'s in vibration. Everything else is the same, and $n_{bent} = 6 - 1 - 1 - 1 = 3$. Thus, $p_{bent} = 0.008$. Now, Table 15.4 shows that the observed steric factor is 0.011. This is closer to p_{bent} than p_{linear}, and we predict that the transition state is bent.

$\underline{NO + O_3}$: In a bent transition state there are five atoms and 15 total d.f.'s. As always for a bent entity, there are three translational d.f.'s and three rotational d.f.'s. This leaves nine d.f.'s in vibration. NO is necessarily linear and has three translational d.f.'s and two rotational d.f.'s, leaving only one vibrational d.f. O_3, which is bent (remember VSEPR), has nine total d.f.'s. Therefore, it has three vibrational degrees of freedom, so $n_{bent} = 9 - 1 - 3 - 1 = 4$. Thus, $p_{bent} = 0.002$. This is relatively close to the observed 0.005.

$\underline{C_2H_4 + H_2}$: In a bent transition state there are eight atoms and 24 total d.f.'s. As always for a bent entity, there are three translational d.f.'s and three rotational d.f.'s. This leaves 18 d.f.'s in vibration in the transition state. C_2H_4 is not linear and has 18 total d.f.'s. So, it must have 12 vibrational d.f.'s. H_2 is necessarily linear and has only one vibrational d.f., so $n_{bent} = 18 - 12 - 1 - 1 = 4$. Thus, $p_{bent} = 0.002$. Now, this value is very far from the experimental value and reveals that this crude model can't handle many atom molecules.

37.   The figures that we need to draw require only a slight modification of those from Exercise 36, all of which, except (c) appear in the Exercise itself. The (c) sketch will look very much like the other unsuccessful tries, except that there will be less "wiggle" (vibrational energy) in the trajectory and less climbing up the barrier hill. This is because in the unsuccessful text figures $E_v > E_a$ and $E_t > E_a$. For part (c) $E_v < E_a$ and $E_t < E_a$. For part (a) translational energy is produced in excess of E_a. Therefore, the barrier for this reverse reaction must be a late barrier. This is the case that is discussed in the problem description. Thus, the figure is like the early barrier ($E_t > E_a$) case of Exercise 36 – just switch the direction of the arrow! For part (b) vibrational energy is produced in excess of E_a. Therefore, the barrier for this reverse reaction must be an early barrier; the late barrier that *consumed* vibrational energy in the forward reaction must become an early barrier in the reverse reaction and *produce* vibrational energy. Thus, the figure is like the late barrier ($E_v > E_a$) case of Exercise 36. Again, just switch the direction of the arrow. Assuming that part (c) is successful, the barrier

must be located in the middle of the trajectory, since neither energy is, by itself, greater than E_a. Also, since the barrier is in the middle, the energy disposal is not preferential, and we predict the amplitude of the vibration will remain the same after surmounting the barrier as it was before the barrier. However, around the barrier the amplitude is lower since the vibrational (and translational) energy is "drained" as the hill is climbed. Less vibrational energy is shown in these types of figures with less oscillatory behavior in the trajectory.

39. Looked at carefully, this problem is deceptively difficult. The simplifying factors are: i) the concentrations are equal; ii) the solution is perfectly mixed. Consider that Table 15.6 provides information for aqueous *bimolecular* reactions. This means that the reaction occurs from the two reactants colliding and reaction rate = $k[H^+][OH^-]$; the reaction is second order. Thus, reaction rate = $-d[H^+]/dt = -d[OH^-]/dt$. Now, Exercise 6 requires the derivation of $t_{1/2}$ for $2HI \rightarrow H_2 + I_2$, the rate law for which is reaction rate = $k[HI]^2 = kc^2$. For Exercise 39 we can write the rate equation in this same manner, that is, reaction rate = kc^2, *because the initial concentrations of the reactants are equal and their concentrations will remain equal throughout since their rate of consumption is the same.* That is, with c = $[H^+] = [OH^-]$ we can write reaction rate = $-dc/dt = kc^2$. The derivation needed for Exercise 6 is identical to this, except there is no factor of 2. For this exercise $t_{1/2} = 1/c_0 k$. Now, you may say that is the exact equation given below Equation 15.11, meaning we just wasted our time. This is simply not the case. Note that the text example for a second order reaction that leads to $t_{1/2} = 1/kc_0$ has only one reactant, as does Exercise 6. Also, the coefficient of the reactant in each of these cases is two, and the writing of

$$\text{reaction rate} = -\frac{1}{2}\frac{dc}{dt} = kc^2$$

is easy. This was NOT the case for this problem. *There are two reactants.* We only derived the simple equation for $t_{1/2}$ because of the above mentioned simplifications. One could simply look below Equation 15.11 and use the half-reaction time equation found there. This gives the correct answer, but, hopefully, we've shown that doing this misses the entire reason that the equation works. Oh, by the way. $t_{1/2} = 1/[(1.35 \times 10^{11}$ L/mol s$)0.10$ M$]$ = 7.4×10^{-11} s = 74 ps.

41. In this Exercise $t_{1/2}$ refers to the time that it takes to get half-way to *equilibrium*. This dissociation always lies far to the left. The Principle of Detailed Balance tells us that $K = k/k'$, where k' is the diffusion-limited rate constant for the reverse reaction. Equation 15.57 gives

$$k_D = k' = 4\pi(5.0 \times 10^{-8} \text{ cm})\left(\frac{(9.31 + 1.09) \times 10^{-5} \text{ cm}^2}{s}\right) = \frac{6.53 \times 10^{-11} \text{ cm}^3}{\text{molecule s}} = \frac{3.94 \times 10^{10} \text{ L}}{\text{mol s}}$$

Note that the unit on k' comes as no surprise since Equation 15.57 is for k_D, which is a *bimolecular* rate constant. We find that $k = (3.94 \times 10^{10}$ L/mol s$)(1.8 \times 10^{-5}$ mol/L$) = 7.1 \times 10^5$/s. The "unit" of K results from the fact that the standard state concentration is 1 mol/L and the equilibrium equation for this ionization.

Now, we have gone to great lengths to keep track of the coefficients in our rate law expressions. (See Exercises 6 and 39.) To that end, it must be noted that, as far as we know, ionization is an elementary (unimolecular) process. So, rate = $k[CH_3CH_2COOH]$. Therefore, the stoichiometry that we have here is identical to what we have in Equation 15.1. Thus, we can use Equation 15.8 for $t_{1/2}$, which gives $t_{1/2} = \ln(2/(7.1 \times 10^5/s)) = 9.8 \times 10^{-7}$ s.

43. P is formed in only the last step. Thus,

$$\text{rate}_{\text{TST}} = \frac{d[\text{P}]}{dt} = k_{\text{TST}}[(\text{A}\cdots\text{B})]$$

For the other two steps

$$\text{rate}_{\text{D}} = -\frac{d[\text{A}]}{dt} = -\frac{d[\text{B}]}{dt} = k_{\text{D}}[\text{A}][\text{B}] \qquad \text{and}$$

$$\text{rate}_{\text{D}'} = -\frac{d[(\text{A}\cdots\text{B})]}{dt} = k_{\text{D}'}[(\text{A}\cdots\text{B})]$$

The caged complex is present in all three steps, which leads to in the SSA

$$\frac{d[(\text{A}\cdots\text{B})]}{dt} = k_{\text{D}}[\text{A}][\text{B}] - k_{\text{D}'}[(\text{A}\cdots\text{B})] - k_{\text{TST}}[(\text{A}\cdots\text{B})] \approx 0$$

$$\Rightarrow [(\text{A}\cdots\text{B})] = \frac{k_{\text{D}}[\text{A}][\text{B}]}{k_{\text{D}'} + k_{\text{TST}}}$$

Now, substituting this into $d[\text{P}]/dt$ yields

$$\frac{d[\text{P}]}{dt} = \frac{k_{\text{D}}k_{\text{TST}}[\text{A}][\text{B}]}{k_{\text{D}'} + k_{\text{TST}}}$$

If step 2 is faster than diffusion, then $k_{\text{TST}} \gg k_{\text{D}} \approx k_{\text{D}'}$, and the denominator becomes k_{TST}, which leads to $d[\text{P}]/dt = k_{\text{D}}[\text{A}][\text{B}]$. This is identical to rate_{D} that we wrote above and is the diffusion rate. If step 2 limits the rate, then $k_{\text{D}'} \gg k_{\text{TST}}$, which leads to

$$\frac{d[\text{P}]}{dt} = \frac{k_{\text{D}}k_{\text{TST}}[\text{A}][\text{B}]}{k_{\text{D}'}} = k_{\text{TST}}[\text{A}][\text{B}]$$

In the last step, we used the fact that $k_{\text{D}} \approx k_{\text{D}'}$.

45. Equation 15.60 is derived, of course, from Equation 15.58. In the small-[S] limit, we can neglect the $k_1[\text{S}]$ term in the denominator of Equation 15.58, giving

$$\frac{d[\text{P}]}{dt} \approx \frac{k_1 k_2 [\text{E}]_0 [\text{S}]}{k_{-1} + k_2} = \frac{k_1 [\text{E}]_0 [\text{S}]}{\dfrac{k_{-1}}{k_2} + 1}$$

As mentioned in the text, this is first order in [S]. Now, the first step for this mechanism can be written as E + S $\rightarrow$ ES with the rate expression $\text{rate}_1 = k_1[\text{E}][\text{S}]$. We have to show that the above expression for $d[\text{P}]/dt$ reduces to this in the small [S] limit. First, if [S] is small, then E remains essentially constant, and $[\text{E}]_0 \approx [\text{E}]$. Second, k_{-1} will be small in this limit, and the first term in the denominator of our above $d[\text{P}]/dt$ expression goes to zero. Therefore, we have shown that Equation 15.58 (and therefore Equation 15.60) reduces to the rate that results if the first step is the rate-limiting step.

For large [S] we can ignore the k_{-1} and k_2 terms in the denominator of Equation 15.58. By sight, then, you should be able to reduce $d[\text{P}]/dt$ to $k_2[\text{E}]_0$. Once again, though, we are to show that the reduction of Equation 15.58 (or 15.60) in the limit of large [S] gives a rate that is equal to the rate that results if the k_2 step in this mechanism is the rate-limiting step. Now, $\text{rate}_2 = k_2[\text{ES}]$. Neglecting the k_{-1} and k_2 terms in the denominator of [ES] gives

$$\text{rate}_2 = k_2 \left(\frac{k_1 [\text{E}]_0 [\text{S}]}{k_1 [\text{S}]} \right) = k_2 [\text{E}]_0$$

This is identical to the rate that is predicted from Equation 15.58 in the limit of large[S]. Further, it is a constant and independent of [S], as promised in the text.

16
The Nucleus

16 The Nucleus

<u>Your Chapter 16 GOALS:</u>

- Estimate the radius of an atom's nucleus
- Quantitatively relate the binding energy to the mass defect
- Define the strong force
- Determine the binding energy per nucleon and discuss iron's special role in this calculation
- Define activity and specific activity
- Predict decay modes based on an isotope's relation to the band of stability
- Calculate Δm for the four common decay modes discussed in Section 16.2
- Apply kinetics equations to nuclear decay to calculate half-lives and ratios of isotopes in a decaying sample
- Apply the Gamov model to nuclear decay
- Write balanced equations for nuclear reactions
- Calculate and discuss the relevance of the Coulomb barrier to nuclear processes
- Detail the fission of ^{235}U
- Explain isotopic labeling, nuclear medicine, NMR, radioactive dating, and nuclear power plant operation in terms of nuclear processes

<u>Chapter 16 KEY EQUATIONS:</u>

- 16.1, 16.2, 16.5, 16.10, 16.13, 16.14, 16.16, 16.18, 16.19, 16.23, 16.31

<u>Overview</u>

The chemistry of electron transfer or motion has been unwittingly observed and performed for, at least, tens of thousands of years by man. Of course, the *chemistry* of the situation has been understood for very little of that time. The chemistry of the nucleus has been rarely seen or encountered (sunshine and starlight being major exceptions) and was truly inconceivable until the work of Rutherford, Geiger, and Marsden (see Chapter 1), yet the power of nuclear chemistry was soon displayed on those two momentous summer days. Tragedies preceded by tragedies – such is human history. Nuclear chemistry or processes have not been observed, wittingly or not, for the vast majority of human history because the energies are so difficult to release, yet so monstrous. It is a new force, the strong force, that harnesses this stored energy by "turning on" at short distances, thereby keeping nuclei together despite the *very* close proximity of protons. As indicated by Equation 16.1, the nucleus is extremely small. Since the nucleus is still, of course, subject to quantum effects, we can apply the Particle in a Box model to nuclear distances (*a* in Equation B.8) to show that the energy level spacing is very large (see Exercise 2) – *much* larger than the energy spacing between electronic energy levels and larger still than the spacing between rotational and vibrational levels. Equation 16.2 indicates this with the presence of the MeV unit. Thus, thermal agitation or chemical energies, typically hundreds of kcal/mol at the most, are utterly incapable of exciting nuclei to higher energy levels. For these reasons, man has seen green grass, grown food, and started fires – all electronic processes – and generally not seen nuclear processes. That sunrises may endure, we hope that future nuclear processes are confined to isotopic labeling, medicinal purposes, spectroscopy, and power plants. See Section 16.5.

The afore mentioned strong force is qualitatively presented in Chapter 16. The quantitative measure of "bonding" in the nucleus is explored in Exercises 5, 6, and 7 with the quantity BE/A. The larger the average binding energy per nucleon, the greater the stability of the nucleus. For Exercises 6 and 7 the number of "bonds" is given by the formula presented in Exercise 26 of

Chapter 4: $N(N-1)$. Exercises 10 through 17 are relatively simple applications of the decay processes given by Equations 16.8 through 16.16. Those processes for which $\Delta m < 0$ are spontaneous. Exercise 18 is a great illustration of the kinetics vs. thermodynamics theme that is present in Chapter 15. Equation 16.2 gives $\Delta E = -0.954$ MeV, indicating a spontaneous reaction. Based solely on this, ^{197}Au should decay by α emission. Equation 16.23 indicates, however, that a large Coulomb barrier exists because the decay products, ^{193}Ir and ^{4}He nuclei, are positively charged. With respect to Figure 16.7, the decay products are trapped to the left of the Coulomb barrier, allowing the ^{197}Au nucleus to remain intact for very long periods of time. In other words, the decay rate, or reaction rate, is very slow. This is entirely analogous to the situation for Equation 1.1. Exercise 19 is also noteworthy. In the preceding paragraph we mentioned that the spacing between nuclear energy levels is very large. Figure 16.5 examines this spacing in excited ^{60}Ni, one of the products of ^{60}Co decay. The γ rays shown in this Figure are emitted when the ^{60}Ni nucleus drops from its second excited state to its first excited state, followed by another jump to the ground state. Figure 2.3 clearly shows that these transitions involve much greater energies than electronic transitions, which typically appear in the visible to UV region of the spectrum. (X-ray emission is possible for an electronic process, however, if an electron makes a large enough jump when filling a vacancy deep within the electron core.)

If you did not master rate processes in Chapter 15, Chapter 16 offers a second chance. The good news is that nuclear decay processes are strictly first order at normal temperatures. Thus, Equations 16.18 and 16.19 are identical in form to Equations 15.7 and 15.8. The SI unit of activity is the becquerel Bq, where 1 Bq = 1 decay/sec. Thus, 1 Ci = 3.7×10^{10} Bq. Note that the becquerel is a small unit (just as the pascal is), which is why many prefer the use of the curie (and atmospheres), as done in *University Chemistry*. Exercise 22 gives you exposure to the units and terms of nuclear decay, making it an important Exercise. Part (a) follows the ^{238}U example of Section 16.3, and for part (c) use the integrated form of the rate law, Equation 16.19, due to the high activity of ^{82}Sr. For Exercise 24 note that $-dN/dt$ equals the specific activity, 6.89×10^{-12} Ci = 0.255 Bq, leading to the number of ^{14}C atoms in a 1 g sample of natural carbon via Equation 16.18. Exercise 25 deals with radiation dose, for which the SI unit is the gray Gy. This is not commonly used, however, a fact that might be known to you because you probably *have* heard of the terms rad and rem for radiation dose but not the gray. When solving Exercise 26, use the equation just below Equation 16.31 with $N_{D0} = 0$. You don't need to convert the given masses to the actual numbers of each isotope because $N \propto n$. Expressing the masses as moles is sufficient. For the solution to Exercise 34 you must determine the energy released per decay of an ^{131}I isotope from Δm and calculate the number of grams that must decay to yield 0.1 rad (= 0.001 Gy).

For nuclear reactions the colliding nuclei approach the Coulomb barrier from the right in Figure 16.7, whereas decaying nuclei are trying to escape to the left in this Figure. (Left/right applies only to the schematics of Figure 16.7. No actual direction is meant.) If the colliding species can surmount the barrier, the strong force takes over and holds the protons of the new nucleus together. The fraction of molecules with the energy to succeed at this is given by the Boltzmann factor, as shown in Exercise 38. See also the text after Example 15.5. Even at the huge temperature of 10^8 K, a Boltzmann factor of $\exp(-280.3)$ is obtained. (As with Exercise 35 of Chapter 9, we do not want you to report this value as zero – even if your calculator tells you that! The methods shown there allow you to evaluate this very small number in order to obtain the value that is found in the text answer key.) That so few nuclei have sufficient energy to surmount the Coulomb barrier at such a large temperature without tunneling is truly amazing and indicates the gargantuan energies at play in nuclear reactions. Further, this miniscule ratio indicates that the rate would be vanishingly small, yet this reaction occurs due to the resonance of ^{8}Be and ^{4}He binding energies with the second excited state of ^{12}C. Exercise 39 affords you the

chance to incorporate your Chapter 15 experience with Chapter 16 via the Steady State Approximation. For Exercise 40 use the mass of ^{1}H not ^{1}p. In Exercise 44(a) V_c refers to the reactants since there can be no Coulomb barrier for the products, as neutrons have no charge. For part (b) the ^{9}Be target has no kinetic energy, giving products of kinetic energy $7.8 + 5.7 = 13.5$ MeV. The thermoneutral designation for the reaction of Exercise 50 indicates that $\Delta E = 0$ and, from Equation 16.2, that $\Delta m = 0$, allowing simple conservation of mass principles to be used.

High resolution NMR gives a pleasing spectrum. See Figure 16.14(b). An IR spectrum of any typical organic molecule is simply more cluttered with broader peaks due to the large number of allowed transitions. See Figure 8.9. NMR spectroscopy monitors a smaller number of transitions than IR because NMR "looks for" nuclei in the "spin up" or "spin down" state. This greatly simplifies the signals that are present, as the nuclei only have two choices. Exercise 54 examines how the sharp NMR spectrum can be made even sharper by increasing the transition frequency. The significant figure hawks might shudder at the results in the text answer key, but the effect is absolutely real and verifiable. There is no question that lowering the temperature and increasing the transition frequency improve the resolution of an NMR spectrum. Boltzmann, indeed, placed v and T in the correct place in his "factor." (There is little that Boltzmann didn't do right, actually.)

16 The Nucleus

1. Density is mass/unit volume. For $A = 1$ Equation 16.1 gives $r_{proton} = 1.33$ fm. Assuming a spherical volume for a proton, whose mass is given in Table 16.1, we find

$$\text{density} = \frac{1.0 \text{ amu}}{\frac{4}{3}\pi (1.33 \text{ fm})^3} = \frac{0.1015 \text{ amu}}{\text{fm}^3}$$

With 1 g = 6.0221×10^{23} amu we easily convert to 1.69×10^{14} g/cm^3. Lastly, with 1 metric ton = 1×10^6 g we get 1.69×10^8 metric tons/cm^3. This is a rough estimate since Equation 16.1 is not so accurate for small nuclei and because nuclei other than ^{1}H have protons and neutrons, which have (slightly) different masses.

3. The spacing between adjacent rotational levels j and $j - 1$ is given by Equation 8.14:

$$\frac{\Delta \varepsilon_{j,j-1}}{hc} = 2Bj = 2B, \text{ since } j = 1$$

B is the rotational constant and can be calculated by the engineering formula given in the solution to Example 8.1. Using the units appropriate for this formula,

$$B(\text{cm}^{-1}) = \frac{4.106^2}{(1.5 \times 10^{-5} \text{ Å})^2 (0.5039851 \text{ amu})} = 1.49 \times 10^{11} \text{ cm}^{-1} = 1.49 \times 10^{13} \text{ m}^{-1}$$

Equation 8.14 then gives

$$\Delta \varepsilon_{j,j-1} = 2hcB = 2(6.6261 \times 10^{-34} \text{ J s})(2.9979 \times 10^8 \text{ m/s})(1.49 \times 10^{13} \text{ m}^{-1})$$

$$= 5.92 \times 10^{-12} \text{ J} = 36.9 \text{ MeV} \approx 37 \text{ MeV}$$

Now, as in Exercise 2, this energy is much greater than the binding energy of the deuteron, which is only 2.224 MeV. Absorption of 37 MeV will, therefore, destroy the deuteron. Thus, based on the rigid-rotor model, we predict that there are no bound rotationally excited states for the deuteron.

5. Equation 16.5 gives BE/A = {[2(1.00727647 + 0.00054858) + 1.00866490 – 3.016029] × 931.5}/3 = 2.57 MeV. As shown at the end of Section 16.1 for ^{56}Fe, with the deuteron as our yardstick, there are 2.57 MeV/(1.112 MeV/bond) = 2.3 ≈ 2 bonds to nearest neighbors. If *each* nucleon makes two bonds, the arrangement must be triangular. Table 16.3 reveals that ^{3}He and ^{1}n have magnetic moments that are of the same sign and almost the same magnitude. This near equality should be the case if the spins of the two protons cancel since only the neutron will be left to contribute to the magnetic moment of ^{3}He.

7. Equation 16.5 gives BE/A = {[6(1.00727647 + 0.00054858) + 6(1.00866490)12]931.5}/12 = 92.162 MeV/12 = 7.68 MeV. If all nucleons could bond to all others, the number of bonds is given by 12(12 – 1)/2 = 66. (See Exercise 26, Chapter 4.) Using the BE that we just determined, this indicates a nucleon "bond energy" of 92.162 MeV/66 = 1.40 MeV/bond. This is less than the "bond energy" in the deuteron. If, however, the nucleon "bond strength" in ^{12}C is the same as in the deuteron, we predict that there are only 92.162 MeV/(2.224 MeV/ bond) = 41.4 bonds ≈ 40 bonds between the nucleons.

9. The math of this problem is easy. The concept of particles communicating force is another matter. Rearranging the given equation yields $m = h/2\pi ac$. Substitution gives

$$m = \frac{6.6261 \times 10^{-34} \text{ J s}}{2\pi (1.5 \times 10^{-15} \text{ m})(2.9979 \times 10^8 \text{ m/s})}$$

$$= 2.35 \times 10^{-28} \text{ kg} \left(\frac{6.0221 \times 10^{26} \text{ amu}}{1 \text{ kg}} \right) = 0.14 \text{ amu}$$

The strong force acts over a very short distance. What about the Coulomb force? From $a = h/2\pi mc$, a mass of zero gives an undefined expression, but in the limit of $m \to 0$, $a \to \infty$. Thus, the Coulomb force never ceases to act at any distance; its range is infinite. At large distances, however, it is too weak to exert any significant force.

11. As shown in Example 16.1 for isotopes with $Z < 84$, neutron-poor isotopes are expected to undergo β^+-decay or EC. Neutron-rich isotopes are expected to undergo β-decay. Example 16.1 describes what is meant by rich and poor. Now, of the three isotopes with $Z < 84$, ^{125}I and ^{11}C are neutron-poor, and ^{3}H is neutron-rich. ^{228}Th has $Z > 83$, and we expect α-decay. See the text key for the balanced nuclear equations. To resolve any ambiguity we check Table 16.2. There we find that ^{3}H does in fact undergo β-decay, ^{11}C undergoes β^+-decay, and ^{125}I undergoes EC and γ-ray emission. ^{228}Th, however, is not listed. It is important to note that, while α-decay is seen for most isotopes with $Z > 83$, it is not the only means of decay. Note Figure 16.8 and ^{234}Th in Table 16.2. To be certain, always look up the "correct" decay mode in a table. Lastly, positron emission tomography (PET) needs, oddly enough, positrons. β^+-decay produces positrons, and, therefore, the neutron poor ^{11}C should be useful for PET.

13. The balanced nuclear equations are:

$$\text{EC}: {}^{82}_{38}\text{Sr} + {}^{0}_{-1}\text{e} \to {}^{82}_{37}\text{Rb} \qquad \beta^+: {}^{82}_{38}\text{Sr} \to {}^{82}_{37}\text{Rb} + {}^{0}_{1}\bar{\text{e}} \qquad \alpha: {}^{82}_{38}\text{Sr} \to {}^{78}_{36}\text{Kr} + {}^{4}_{2}\text{He}$$

Using Equations 16.14, 16.13, and 16.16, respectively, the mass changes are $\Delta m_{\text{EC}} = (81.918195 - 81.918414)\text{amu} = -0.000219$ amu, $\Delta m_{\beta^+} = (81.918195 - 81.918414)\text{amu} + 2(0.00054858 \text{ amu}) = 0.000878$ amu, and $\Delta m_\alpha = (77.920396 + 4.002602 - 81.918414)\text{amu} = 0.004584$ amu. Only EC is feasible because it is the only mode with $\Delta m < 0$.

15. See the text key for the nuclear equations for the β-decay and the EC. The β^+-decay nuclear equation is

$$ {}^{40}_{19}\text{K} \to {}^{40}_{18}\text{Ar} + {}^{0}_{1}\bar{\text{e}} $$

The mass changes, using Equations 16.10, 16.14, and 16.13, respectively, are $\Delta m_\beta = (39.962591 - 39.963999)\text{amu} = -0.001408$ amu, $\Delta m_{\text{EC}} = (39.962384 - 39.963999)\text{amu} = -0.001615$ amu, and $\Delta m_{\beta^+} = (39.962384 - 39.963999)\text{amu} + 2(0.00054858 \text{ amu}) = -0.000518$ amu. Since all three modes have negative Δm's, all three are spontaneous.

17. We begin, as usual, with the nuclear equation: ${}^{214}_{84}\text{Po} \to {}^{210}_{82}\text{Pb} + {}^{4}_{2}\text{He}$. Equation 16.16 then gives $\Delta m = (209.984163 + 4.002602 - 213.995176)\text{amu} = -0.008411$ amu. This leads to an energy of -7.835 MeV, which, in this problem, we assume appears as the kinetic energy of the α particle.

19. The energies are 1.88×10^{-13} J and 2.13×10^{-13} J. The wavelengths come from $\lambda = hc/\Delta E$. They are 1.06×10^{-12} m and 9.33×10^{-13} m. (0.0106 Å and 0.00933 Å) The γ-rays account for $(1.17 + 1.33)\text{MeV} = 2.50$ MeV. That leaves 0.32 MeV as the *maximum* kinetic energy the β particle can possess. Therefore, $K_\beta \leq 0.32$ MeV.

21. From Equation 16.19, $N = N_0 e^{-kt}$. Differentiating this with respect to t gives $dN/dt = -kN_0 e^{-kt} = -kN$, the confirmation of Equation 16.18. [Note: $d(e^u)/dx = e^u(du/dx)$]. Solving for N gives $N = -(dN/dt)/k$. We could also get this by solving Equation 16.18. Now, the initial counting rate must, likewise, give $(dN/dt)_0 = -kN_0$, or $N_0 = -(dN/dt)_0/k$. Placing these two expressions into Equation 16.19 gives $-(dN/dt)/k = -[(dN/dt)_0/k]e^{-kt}$, or $(dN/dt) = (dN/dt)_0 e^{-kt}$.

23. Table 16.2 reveals that this isotope is quite unstable. Remember that the "chemically useful" designation for an isotope is reserved for isotopes whose half-life is no shorter than about one minute. As needed for Exercise 22, we must use the integrated form of the rate law to find the time for the decay rate to decrease. The initial decay rate is

$$-\frac{dN}{dt} = \frac{N \ln 2}{t_{1/2}} = \frac{70. \,\text{amol}\left(\dfrac{1 \,\text{mol}}{1 \times 10^{18} \,\text{amol}}\right)\left(\dfrac{6.0221 \times 10^{23}}{\text{mol}}\right)\ln 2}{1.273 \,\text{min}\left(\dfrac{60 \,\text{s}}{1 \,\text{min}}\right)} = \frac{3.826 \times 10^{5} \,\text{deays}}{\text{s}}$$

We solve the integrated rate law for t by first dividing each side by N_0 and then taking ln of both sides. This yields the equation given in Example 16.2. As always, N is proportional to the activity, and

$$t = \frac{-1.273 \,\text{min}\left(\dfrac{60 \,\text{s}}{1 \,\text{min}}\right)\ln\left(\dfrac{5000}{3.826 \times 10^{5}}\right)}{\ln 2} = 480 \,\text{s} \approx 8.0 \,\text{minutes}$$

25. The nuclear equation for this process is in the text key for Exercise 15. It was also determined there that $\Delta m = -0.001408$ amu. Thus, $\Delta E = -1.312$ MeV $= -2.101 \times 10^{-13}$ J. The kinetic energy, then, of each α particle is 2.101×10^{-13} J. This can be multiplied by the activity in Bq to give the energy received per second. The activity is, as usual, $N\ln 2/t_{1/2}$. To get N, we first need the mass of ^{40}K present in the body. The given conditions indicate that there are 50 kg(0.0035)(0.00012) = 0.000021 kg = 0.021 g of ^{40}K in the body. The activity is

$$-\frac{dN}{dt} = \frac{0.021 \,\text{g}\left(\dfrac{1 \,\text{mol}}{39.963707 \,\text{g}}\right)\left(\dfrac{6.0221 \times 10^{23}}{1 \,\text{mol}}\right)\ln 2}{1.28 \times 10^{9} \,\text{y}\left(\dfrac{3.16 \times 10^{7} \,\text{s}}{\text{y}}\right)} = 5423 \,\text{Bq}$$

Multiplying these two numbers reveals that a 50 kg mass receives

$$\text{dose} = \frac{5423 \,\text{decays}}{\text{s}}\left(\frac{2.101 \times 10^{-13} \,\text{J}}{\text{decay}}\right)\left(\frac{3.16 \times 10^{7} \,\text{s}}{\text{y}}\right) = \frac{0.0360 \,\text{J}}{\text{year}}$$

The SI unit of radiation dose is the gray, Gy. It is defined as joules of energy absorbed per kilogram of absorbing material. The rad is defined in terms of ergs and grams, and 1 Gy = 100 rad. This 0.0360 J is hitting a 50 kg mass. Therefore, the radiation dose per year is (0.0360 J/year)/50 kg = 0.000720 Gy/year = 0.0720 rad/year. Finally, at the beginning of the *Nuclear medicine* subsection of Section 16.5, we see for β and γ radiation that 1 rem = 1 rad, meaning that the radiation dose per year in rem is 0.0720 rem/year.

27. This solution is very similar to that shown in Example 16.2. The integrated rate law is solved for t to give

$$t = \frac{-t_{1/2}\ln\left(\dfrac{N}{N_0}\right)}{\ln 2} = \frac{-(4200000 \,\text{y})\ln(1 \times 10^{-12})}{\ln 2} = 1.7 \times 10^{8} \,\text{years.}$$

Since N is proportional to the number of moles, we use the given mol ratio (i.e., mol fraction) for the ratio of N to N_0. Note this is the *minimum* age that we can assign to the solar system with the information in this problem. We are limited by the sensitivity of present analytical instrumentation. If the *actual* value for the mol fraction of ^{98}Tc remaining

is less than 10^{-12}, then the value of t given by the above equation would be larger, meaning an older solar system.

29. The nuclear equations are

$$^{232}_{90}\text{Th} \rightarrow ^{228}_{88}\text{Ra} + ^4_2\text{He} \quad \text{and} \quad ^{230}_{90}\text{Th} \rightarrow ^{226}_{88}\text{Ra} + ^4_2\text{He} \quad \text{and} \quad ^{228}_{90}\text{Th} \rightarrow ^{224}_{88}\text{Ra} + ^4_2\text{He}$$

Equation 16.16 leads to Δm's of -0.004388 amu, -0.005123 amu, and -0.005927 amu for ^{232}Th, ^{230}Th, and ^{228}Th, respectively. Now, the more negative Δm gets, the larger is the energy change associated with the decay. This larger amount of energy appears in the products and can assist the α particle in tunneling through the Coulomb barrier. This leads to a shorter half-life and the order shown in the text key. Comparison of our prediction with the measured values for the half-lives, which are listed in the text key, verifies our ordering.

In order to sketch the figure, we need to determine the ΔE's for the three decay processes and the V_c's. From the above Δm's the ΔE's are -4.087 MeV, -4.772 MeV, and -5.521 MeV. V_c is the barrier through which the products of the decay must tunnel. Since this formula contains the radius of the daughter atom, there will be a slightly different Coulomb barrier for each case. For the ^{228}Ra case:

$$V_c = \frac{1.4400(2)(88)}{1.33(\sqrt[3]{228} + \sqrt[3]{4})} = 24.8 \text{ MeV}$$

For ^{226}Ra and ^{224}Ra, $V_c = 24.8$ and 24.9 MeV, respectively. Within the accuracy of our sketch, V_c is approximately equal for all three isotopes. With this approximation the sketch is very similar to Figure 16.7, except, of course, for the different values. The ^{228}Th isotope is the least stable and will have the least tunneling to do, and the ^{232}Th isotope has the most tunneling to do. This should be clear from your sketch.

31. a. Figure 16.8 directly reveals that there are 7α and 6β emissions in this chain. However, there is an easy analytical solution to this problem, as well, *due to the assumption that only α and β decays are possible.* Now, the change in mass number, ΔA, must be divisible by four if α decay is to be possible. For the decay from ^{238}U to ^{210}P, the change in mass number is 28. This means that there must be seven α decays to get $\Delta A = 28$. However, this would mean that the change in atomic number, Z, would be $2(7) = 14$. For ^{238}U to ^{210}P, though, $\Delta Z = 8$. The last piece of the solution relies on the fact that β decays increase Z and, thereby, decrease ΔZ. Therefore, there must be six β decays to get $\Delta Z = 8$.

    b. The ratio N/N_0 from Equation 16.19 is the fraction of atoms in a radioactive sample that has not decayed after a given time t. The fraction of the sample that is still ^{210}Po after 5 years is

    $$\frac{N}{N_0} = \exp\left(-(\ln 2)\left[\frac{5 \text{ y}}{138 \text{ d}/(365 \text{ d/y})}\right]\right) = 0.000104$$

    The fraction of the sample, then, that is ^{206}Pb is $1 - 0.000104 = 0.999896$.

    c. ^{210}Po decays by alpha emission, which explains the presence of the helium gas. $0.999896(0.200 \text{ g}) \approx 0.200$ g of this sample will be converted to ^{206}Pb and He gas. Thus, the moles of He gas formed is given by

    $$0.200 \text{ g } ^{210}\text{Po}\left(\frac{1 \text{ mol}}{209.98 \text{ g}}\right)\left(\frac{1 \text{ mol He}}{1 \text{ mol } ^{210}\text{Po}}\right) = 0.0009525 \text{ mol He}$$

    Lastly, Equation 9.13 gives $P = (0.0009525 \text{ mol})R(298 \text{ K})/0.100 \text{ L} = 0.233$ atm.

33. The activity of the sample is $100 \text{ mCi} = 0.100 \text{ Ci} = 3.70 \times 10^9 \text{ Bq}$. Equation 16.18, written for the initial rate of decay, gives

$$N_0 = \dfrac{-\left(\dfrac{dN}{dt}\right)_0 t_{\frac{1}{2}}}{\ln 2} = \dfrac{\left(\dfrac{3.70 \times 10^9 \text{ decays}}{\text{sec}}\right)163\text{ days}\left(\dfrac{86{,}400 \text{ sec}}{\text{day}}\right)}{\ln 2} = 7.52 \times 10^{16} \text{ atoms}$$

(Remember that $-dN/dt$ is a positive number, as in the text example for ^{238}U in Section 16.3.) This is now a problem like we solved in Chapter 1. Using the isotope mass from Table 16.1, this many atoms of ^{242}Cm has a mass of 3.02×10^{-5} grams $= 30.2$ µg.

35. The ratio N/N_0 from Equation 16.19 is the fraction of atoms in a radioactive sample that has not decayed after a given time t. The half-life is given in Table 16.2 as 29 years. Thus, after 50 years,

$$\frac{N}{N_0} = \exp\left[-(\ln 2)\left(\frac{50.}{29}\right)\right] = 0.30$$

meaning that 30. % of the ^{90}Sr is still there.

37. The two equations add very simply, and the net nuclear equation is given in the text key. Since the amu scale is relative to ^{12}C, it makes for an easier calculation, and the γ-ray does not affect Δm, which explains why it is sometimes omitted from the nuclear equation. Thus, $\Delta m = 12 - 3(4.002602) = -0.007806$ amu, and $\Delta E = -7.271$ MeV $\approx -1.165 \times 10^{-12}$ J. If all of this energy appears as a γ-ray, then the minimum wavelength of the radiation is given by $\lambda = hc/\Delta E = 1.705 \times 10^{-13}$ m $= 0.001705$ Å.

39. a. The nuclear equations are given in the text key. The overall addition is very easy. As was done with Equations 16.3 and 16.4, however, the proton is written as the entire isotope. If for no other reason, this is done because the other atoms are also written as the complete isotope (^{12}C, ^{13}N, etc). Writing ^{1}p as the *only* nucleus would make the equations impossible to balance. The Δm's are (recall, γ-rays don't affect Δm): $\Delta m_1 = 13.005738 - (12 + 1.00782505) = -0.002087$ amu; $\Delta m_2 = 13.003355 - 13.005738 + 2(0.00054858) = -0.001286$ amu; $\Delta m_3 = 14.003074 - 13.003355 - 1.00782505 = -0.008106$ amu. These Δm's give energy changes of -1.944 MeV, -1.198 MeV, and -7.551 MeV. The energy change for the overall reaction is just the sum of the individual steps, giving $\Delta E_{\text{tot}} = -10.693$ MeV.

 b. There are three steps to this mechanism. The rate constants have been indicated in the text key. Chapter 15 contains many examples of the use of the SSA. Here, we state the conditions for the intermediates:

$$\frac{d[^{13}_7\text{N}]}{dt} = k_1[^{12}_6\text{C}][^1_1\text{H}] - k_2[^{13}_7\text{N}] \quad \text{and} \quad \frac{d[^{13}_6\text{C}]}{dt} = k_2[^{13}_7\text{N}] - k_3[^{13}_6\text{C}][^1_1\text{H}]$$

Each of these expressions is approximately equal to zero in the SSA, and the first expression gives

$$k_1[^{12}_6\text{C}][^1_1\text{H}] \approx k_2[^{13}_7\text{N}] \quad \Rightarrow \quad [^{13}_7\text{N}] \approx \frac{k_1[^{12}_6\text{C}][^1_1\text{H}]}{k_2}$$

Substituting this into the second rate expression, set equal to zero, and solving for $[^{13}\text{C}]$ gives

$$k_2\left(\frac{k_1[^{12}_6\text{C}][^1_1\text{H}]}{k_2}\right) \approx k_3[^{13}_6\text{C}][^1_1\text{H}] \quad \Rightarrow \quad [^{13}_6\text{C}] \approx \frac{k_1[^{12}_6\text{C}]}{k_3}$$

The first step in the mechanism is the key step since: i) we are considering *only the forward reactions*; ii) the reactants of steps two and three are intermediates, and the overall rate equation can't contain the intermediates. Lastly, each step in a mechanism is elementary, leading to the overall rate equation found in the text key.

41. The minimum wavelength corresponds to the maximum energy that the γ-ray can possess. This maximum energy would arise from the total energy change of the reaction appearing entirely in the γ-ray. To get ΔE, we begin again with Δm: $\Delta m = 23.985042 - 2(12) = -0.014958$ amu. Thus, $\Delta E = -13.93$ MeV $\approx -2.232 \times 10^{-12}$ J. Therefore, $\lambda = hc/2.232 \times 10^{-12}$ J $= 8.900 \times 10^{-14}$ m $= 0.0008900$ Å.

43. Assuming that all of the translational energy is due to the α particle, $E_t = 7.835$ MeV. Equation 16.1 gives r_{He} and r_N. In the text it is shown the $V_c = 3.8$ MeV. Equation 16.25 then gives

$$\sigma = \pi[1.33(\sqrt[3]{4} + \sqrt[3]{14})]^2\left(1 - \frac{3.8}{7.835}\right) = 45.73 \text{ fm}^2\left(\frac{1 \text{ cm}}{1 \times 10^{13} \text{ fm}}\right)^2\left(\frac{1 \text{ b}}{1 \times 10^{-24} \text{ cm}^2}\right) = 0.46 \text{ b}$$

Now, in Equation 16.24 $dN/dt \propto \sigma$. In the line-of-centers model the nuclear reaction cross section σ becomes that given in Equation 16.25, the only difference being the term $(1 - (V_c/E_t))$. It is this term, then, that is responsible for the reduction of the rate estimate given by Equation 16.24. In this instance its value is 0.51. Reducing a value by 50 %, of course, is no small correction.

45. a. Beginning again with the mass change, $\Delta m = 1.00866490 + 29.978307 - (26.981539 + 4.002602) = 0.002831$ amu, and $\Delta E = 2.637$ MeV. To calculate V_c, note that it applies to the reactants alone since there is no Coulomb barrier for the products. Thus, we need the radius of ^{27}Al and ^{4}He, not ^{30}P and ^{1}n.

The use of Equation 16.1 along with Equation 16.23, easily gives $V_c = 6.14$ MeV, and Equation 16.25 gives

$$\sigma = \pi[1.33(\sqrt[3]{27} + \sqrt[3]{4})\text{fm}]^2\left(1 - \frac{6.14}{7.835}\right) = 0.253 \text{ b}$$

The conversion to barns is shown in Exercise 43. The estimate of the kinetic energy of the products is the difference of the α particle's kinetic energy and ΔE for the reaction. Here, the reaction is endothermic. Some of the α particle's kinetic energy is used to provide the needed energy. (In Exercise 44 the reaction is exothermic. We added the two energies there.) Now, if the target atom is stationary, it has zero kinetic energy, so 7.835 MeV – 2.637 MeV = 5.198 MeV = kinetic energy of products. Note that V_c won't affect the kinetic energy because the neutron product tunnels through the Coulomb barrier.

    b. The reaction that we consider is $^{27}_{13}$Al $+ ^4_2$He $\rightarrow ^{31}_{15}$P. $\Delta m = 30.973762 - (26.981539 + 4.002602) = -0.010379$ amu, leading to $\Delta E = -9.668$ MeV.

47. As with Equations 16.3 and 16.4, the entire isotopic form is used. The nuclear equation is

$$^2_1\text{H} + \gamma \rightarrow ^1_1\text{H} + ^1_0\text{n}$$

$\Delta m = 1.00782505 + 1.00866490 - 2.014102 = 0.002388$ amu, and $\Delta E = 2.224$ MeV $\approx 3.564 \times 10^{-13}$ J. Assuming all of the energy to cause the dissociation is due to the γ-ray, then the wavelength needed is given by $\lambda = hc/\Delta E = 5.574 \times 10^{-13}$ m ≈ 0.005574 Å.

49. a. The nuclear equation is given in the text key. From this $\Delta m = 2(115.914) + 4(1.0086649) - [235.043924 + 1.0086649] = -0.190$ amu, and $\Delta E = -177$ MeV. Now, there is no Coulomb barrier for the reactants. In problems where the reactants both contain positive charge, this is not the case. The neutron, of course, has no charge, and we consider V_c based on the two ^{116}Pd nuclides only. Equation 16.1 gives $r_{Pd} = 6.49$ fm. Equation 16.23 then gives $V_c = 1.4400(46)(46)/[6.49 + 6.49] = 235$ MeV.

    b. Reaction 16.28 gives $\Delta m = 93.938 + 138.908826 + 3(1.0086649) - [235.043924 + 1.0086649] = -0.179768 = -0.180$ amu, and $\Delta E = -167$ MeV. The two nuclei to consider are ^{94}Kr and ^{139}Ba. With $r_{Kr} = 6.05$ fm, and $r_{Ba} = 6.89$ fm, $V_c = 1.4400(36)(56)/12.94 = 224.3$ MeV. Despite the fact that the ^{235}U fission to ^{116}Pd has a greater ΔE and, therefore, is assumed to be more spontaneous, its larger V_c prevents the reaction from occurring at a measurable rate. The second reaction has a smaller barrier, and its occurrence is more likely.

51. From Exercise 49 $\Delta E = -167$ MeV for Equation 16.28. The fuel is ^{235}U, and ΔE/mass = 167 MeV/235.04 amu = 0.711 MeV/amu = 6.86×10^{13} J/kg = 1.64×10^{10} cal/g. These last conversions should be very easy for you. For Equation 16.32 $\Delta m = 3.016029 + 1.0086649 - 2(2.014102) = -0.003510$ amu, and $\Delta E = -3.270$ MeV. The fuel here is ^{2}H, and ΔE/mass = 3.270 MeV/2(2.014102) = 0.8118 MeV/amu = 7.832×10^{13} J/kg = 1.872×10^{10} cal/g. This demonstrates one of the reasons why fusion is the energy hope of the future. It simply is more efficient.

53. Given in the footnote to Table 16.3 is the energy splitting between adjacent magnetic levels: the so-called "spin-up" or "spin down" states. The formula is $\Delta E = (\mu/I\mu_N)\mu_N B$. Using Table 16.3 to find the values for ^{13}C gives
$$\Delta E = (2)(0.702)(5.05 \times 10^{-27} \text{ J})(1 \text{ T}) = 7.09 \times 10^{-27} \text{ J}$$
Note that $1/I = 1/\frac{1}{2} = 2$. Now, since $E = h\nu$, the frequency of the resonance is $\nu = (7.09 \times 10^{-27}$ J)/(6.6261 \times 10^{-34}$ J s) = 1.07×10^7 Hz = 10.7 MHz. ^{19}F is done in the identical manner. We find that $\Delta E = 2.66 \times 10^{-26}$ J and that the frequency of the resonance is 4.01×10^7 Hz = 40.1 MHz.

55. The Bohr postulate is $E = h\nu$, or $\Delta E = h\Delta\nu$. Substituting this into the energy-time uncertainty principle gives
$$\Delta t \geq \frac{h}{2\pi h \Delta\nu} \geq \frac{1}{2\pi \Delta\nu}$$
The definition of chemical shift is
$$\delta = 1 \times 10^6 \left(\frac{\Delta\nu}{\nu_0} \right)$$
If $\Delta\delta = 14$, then $\Delta\nu = 14(1 \times 10^{-6})\nu_0 = 14(1 \times 10^{-6})(5.00 \times 10^8 \text{ Hz}) = 7.00 \times 10^3$ Hz. Substitution into the derived energy-time uncertainty principle gives $\Delta t = 2.3 \times 10^{-5}$ seconds. Any pulses longer than this will result in an energy span that is too small to cover the possible energies, or frequencies, that protons exhibit. Thus, the pulse must be less than or equal to 2.3×10^{-5} seconds.

17

The Transition Metals

17 The Transition Metals

<u>Your Chapter 17 GOALS</u>:

- Note the ionization energy and ionic radii trends of the $3d$ metals and relate them to electronic structure
- Learn the common oxidation states and outline the chemistry for the $3d$ metals
- Draw qualitative absorption spectra for transition metal complexes
- Use MO theory to explain the origins of color in transition metal complexes
- Describe charge transfer transitions
- Use the "Eyeball Spectroscopy" chart to predict the energy level spacing in transition metal complexes
- Name transition metal complexes
- Relate a ligand's structure to its bonding properties
- Apply formation constant K_f principles to complex ion formation
- Define labile
- Use crystal field theory and ligand field theory to rationalize the order of the spectrochemical series and to provide a systematic explanation for the color of transition metal complexes
- Relate "bond strength" of M–L bonds and colors of transition metal complexes to Δ_o, the splitting parameter
- Use kinetic arguments to describe ligand substitution and oxidation-reduction reactions
- Describe the outer-sphere and inner-sphere mechanisms for electron transfer between complexes
- Identify the key aspects of the $4f$ orbital boundary surfaces
- Explain the common oxidation states of the lanthanides with electronic structure arguments

<u>Chapter 17 KEY EQUATIONS</u>:

- 17.2, 17.4, 17.7, 17.11

<u>Overview</u>

Chapter 17 offers a rest from the math-intensive Chapters that comprise the lion's share of *University Chemistry*. It is essentially a qualitative survey of the principles of what is often called *inorganic chemistry*, an area of chemistry that studies mainly the transition metals. As we have argued previously, however, such designations are arbitrary, and chemistry is best viewed as a whole, evidenced in Chapter 17 by the appearance of LCAO-MO theory, electronic transitions, kinetic/rate equations, formation constants, etc.. Hopefully, you will experience the chemistry and colors of the transition metal ions in a laboratory course, as the colors are a far cry from the colorless monotony of much of main group chemistry. (The ligand substitution sequence that is described in Section 17.4 is a classic experiment to perform.) If the material of previous Chapters is to be of use, however, we must be able to explain the colors of transition metals complexes as well as the order of the spectrochemical series, among other things, through the use of the theories we have developed. This is the test of any true theory or model. (Recall the inability of classical physics to describe electrons or even red-hot iron.) What, then, is responsible for the wonderful colors of transition metal complexes or compounds? As demonstrated in Chapter 8, rotational and vibrational energy levels are separated by small energy differences, all of which lead to transitions that occur in the near IR to microwave region. Thus, as usual in chemistry, we turn to the electrons as the source of the chemistry and colors of transition metal complexes.

17 The Transition Metals

A very important point that is driven home by Exercises 1 and 2 is that the electrons in the highest n level are removed first when ions are formed for transition metals. Confusion over this results for some students because the $4s$ electrons fill *before* the $3d$ orbitals when electron configurations are written. The idea that "the last one in is the first one out" is *not* true. Electron configurations are *for isolated, gas phase ions, based on the energy levels shown in Figure 4.8*. Ionization drastically changes the energetic make up for the electrons, and the loss of ns electrons, *which are valence electrons*, is observed. For Exercise 2 Au is anomalous. See Table 4.1. The oxidation number rules for Exercises 3 and 4 are presented near the end of Section 5.6.

As usual the classifications of Exercise 6 are not always easy, especially part (a). Exercise 8 may, perhaps, serve as your introduction to quantitative spectroscopy. However, it is likely that you have seen the Beer-Lambert Law before in a laboratory exercise. The so-called "Spec-20", or any other small, desktop spectrometer, is a staple in undergraduate experiments, and its operation is explained via this Law. The Spec-20 is also used industrial and environmental laboratories; it is by no means just an academic tool. More elaborate spectrometers still rely on the basic absorbance/concentration connection. When solving Exercise 8 note that, as the pathlength grows from 0 to L, the intensity of the light beam that is measured by the detector decreases from I_0 to I. Thus, solve the equation as $-dI/I = \alpha c\, dL$ and integrate the left hand side from I_0 to I and the right hand side from 0 to L. Further, absorbance has no unit. In Exercise 9, assume ionic bonding as in Figure 17.6 even though the compounds in this Exercise are not of the general form MO_4^{n-}. For Exercise 10 we find that in Section 17.2 the series PO_4^{3-}, SO_4^{2-}, ClO_4^- bears a strong resemblance to the series in Figure 17.6, implying that TiO_4^{4-} would resemble SiO_4^{4-}. Also, ΔE_{CT} would be larger as it lies to the left of VO_4^{3-} in Figure 17.6.

Exercise 13 examines the same ligand with a different metal to show that Ni(II)–OH_2 "bonds" are weaker than Co(II)–OH_2 "bonds", whereas Exercise 14 compares the "bonding strength" of two different ligands to the same metal. (See also Exercises 21 and 22.) The ability of one ligand to form stronger M–L "bonds" and to replace another is *the* concept behind the spectrochemical series. The key to Exercise 14 is to see that every NH_3 that is present will bond preferentially to the metal as opposed to a Cl^- ligand, as indicated by Equation 17.4.

The "correctness" of the Crystal Field Model and the Ligand Field Model is secondary to the success of these theories. From our modern view crystal field theory seems unrealistic and not rooted deeply in orbital theory, aside from its inclusion of the d orbitals. At least ligand field theory incorporates the MO theory of Chapter 7, yielding Figure 17.13 for octahedral complexes. Before, however, we banish crystal field theory to the chemistry underworld, note that many chemists still routinely use Lewis structures in research ideas or proposals or in lectures to other advanced scientists who are "in the know", yet Lewis structures are hardly "correct". Research *must* move forward; chemistry must advance. It takes all sizes and shapes of chemists to make the chemistry world go 'round, and many chemists can't worry about the details of a monstrous Hamiltonian. In many ways it is completely irrelevant to them. Lewis structures work in many cases. Likewise, crystal field theory and ligand field theory, for that matter, work in many cases, giving rational explanations for the color, shape, magnetic properties, bonding, and other aspects of transition metal complexes. Useful tools, schematic or not, can never and should never be eliminated from science. The key result of either theory is the *splitting* Δ_o that exists between the d orbitals. When the splitting is large, strong M–L bonds are predicted along with large electron energy level spacing. The opposite is true for small Δ_o's. This parameter also helps to predict and explain magnetic properties via unpaired electrons by comparing its value to the pairing energy between two electrons in the same d orbital. See Exercises 18 through 20. Crystal field theory's very antibonding $x^2 - y^2$ orbital is responsible for Ni^{2+}'s forming square planar complexes almost exclusively. This effect is even more dramatic for Pd^{2+} and Pt^{2+}. Exercise 27

is a thought provoking, difficult Exercise, yet it re-examines many aspects of MO theory and is, therefore, highly useful.

The kinetics Exercises give yet another opportunity to compare and contrast the role of thermodynamics and kinetics in chemical reactions or processes. Complexes can be thermodynamically unstable, yet a low lability, that is, rate of ligand exchange, can keep the complex together for a significant time. The reverse is also true. Exercises 28 and 31 are typical kinetics problems: Express the rate equation based on the slow step of the mechanism and eliminate any intermediates in this rate equation with the SSA. A simple way to solve Exercise 30 is to note that the units of k, the rate constant, indicate a second order reaction (see Exercise 6 of Chapter 15) and use Table 15.2.

Assume thermoneutrality for Exercise 36, as in Exercise 50(b) of Chapter 16. The masses from Table 16.1 can simply be added for $^{235}UF_6$ and $^{238}UF_6$, and these can be used in Equation 9.39. The number of stages then depends on the fractional abundance of ^{235}U. Assume one mole for simplicity.

17 The Transition Metals

1.   The *ns* electrons are the valence electrons for the transition metals, so these are lost first. See the text answer key. Note that Sc^{3+} is sometimes called a d^0 configuration.

3.   See Section 5.6 and the text answer key. Oxygen is -2 in all of these cases. $V(NO_3)_3$ contains the nitrate ion.

5.   a.,b. The general trend for the ionic radius of the 2+ and 3+ ions of the 3*d* metals is that it decreases up until Co, with a larger difference in radius between the early transition metals than in the late ones. Based on Figure 17.4, we pick:

ion pair	smaller ion
Fe^{2+} / Fe^{3+}	Fe^{3+}
Co^{2+} / Fe^{2+}	Co^{2+}
Cr^{3+} / Ti^{3+}	Cr^{3+}

The Fe^{2+}/Fe^{3+} ions have (of course) the same Z, but Fe^{3+} has one less electron. With less electron-electron repulsion (or reduced screening of the outer electrons), Fe^{3+} must be smaller. Figure 17.4 shows that this is clearly the case. The other two cases shown above are more difficult since both Z and the number of electrons changes between ions. Now, there are still two pairs. The case of Pd^{4+} and Rh^{3+} is the reverse of the Fe^{3+}/Fe^{2+} case. Both Pd^{4+} and Rh^{3+} have the electron configuration $[Kr]4d^6$, but Pd has one more proton. For this reason Pd should attract the same number of electrons more strongly and have a smaller radius. Lastly, it is easy to decide between Sc^{3+} and Y^{3+}. *n* is larger for Y, which means that its atomic radius and ionic radius must be larger than Sc's. This trend holds for the main group elements (except Al and Ga) as well as the 3*d* and 4*d* transition metals. You can convince yourself of this by looking at all the values in Figures 4.12 and 4.13. The lanthanide contraction, however, makes such simple predictions impossible when comparing the 4*d* and 5*d* metals.

7.   Assuming these compounds are ionic, each metal is M^{4+}. From Figure 17.2 Mn has the highest IE_3 of these metals. If this holds for IE_4, which it should due to Mn having the highest Z, Mn^{4+} should be the most eager metal ion to get an electron back, making MnO_2 the most likely to be reduced and, therefore, the best oxidizing agent. The predicted order is $MnO_2 > CrO_2 > VO_2 > TiO_2$.

9.   In Figure 17.6 each metal atom is bonded to four oxygen atoms, resulting in the same MO scheme for each compound. The compounds here share no such similarity, but if ionic bonding exists in each case and a charge-transfer transition occurs, we can say that the HOMO to LUMO gap (ΔE_{CT}) must be smallest in V_2O_5 since it has visible absorption. This is attributable to the increasing Z from Sc to Ti to V, resulting in a lowering of the energy of the metal atom AO's (the LUMO's) and, ultimately, a shrinking ΔE_{CT}. Our generic diagram places the bonding MO's, indicated as a single level (not the actual number or exact positions), below the metal *d* AO's, assuming that they remain non-bonding as in Figure 17.6. Sc_2O_3 and TiO_2 must absorb in the UV and, therefore, are colorless.

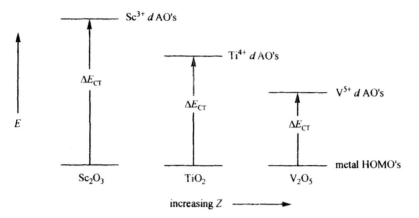

increasing Z ————————▶

11. Whereas Exercise 10 considered the compound to the left of VO_4^{3-} in Figure 17.6, this Exercise considers the compound to the right of MnO_4^-. Since Z is larger for Fe than Mn, the trend of decreasing energy for the $3d$ level and smaller ΔE_{CT} should continue. Thus, the MO picture is just like that for MnO_4^-, except for the difference mentioned above. Assuming a shift in wavelength of roughly 150 nm, then the wavelength of maximum absorbance of FeO_4 would be approximately 710 nm, based on Figure 17.5. We assume 150 nm because the other shifts are about that large as the series moves from VO_4^{3-} to CrO_4^{2-} to MnO_4^-. Table 17.1 shows that the color of such a solution should be nearly green with a hint of blue.

13. a. From Table 17.1 a pale, red-violet solution absorbs light with a λ_{max} somewhere between 500 to 560 nm, and a green solution absorbs light with a λ_{max} at 720 nm. Thus, Co(II) absorbs the shorter wavelength.

    b. Using the relevant λ_{max}'s from part (a) the absorption peak in each case should closely resemble the VO_4^{3-} and CrO_4^{2-} curves in Figure 17.5. See that Figure. The actual λ_{max} of $Co^{2+}(aq)$ is at 510 nm, but with the given color of "red-violet" any λ_{max} between 500 and 560 nm is correct for this sketch. It is impossible to know that "red-violet" means a λ_{max} at 510 nm. The actual curves (~0.1 M) are shown below:

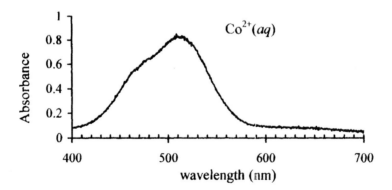

18 The Chemistry of Carbon

<u>Your Chapter 18 GOALS</u>:

- Assess the ability of carbon and its neighbors, boron, nitrogen, and silicon, to make compounds, particularly those in which homopolar bonds are present
- Examine the 3 center 2 electron structure of boron compounds
- Describe the various conformations that hydrocarbons exhibit
- Write skeletal and structural formulas for various hydrocarbons and organic compounds
- Define the various isomers seen in organic compounds
- Describe and list the common reaction types seen in organic chemistry: combustion, halogenation, dehydrogenation, addition, substitution, Diels-Alder, Williams synthesis, and polymerization
- Write mechanisms for multi-step organic reactions
- Identify the role of organic chemistry in biological reactions
- Define the role of proteins, enzymes, and ATP in biological systems
- Describe protein structure
- Outline the process of photosynthesis
- Outline how enzymes work

<u>Chapter 18 KEY EQUATIONS</u>:

- As with much of Chapter 5, the equations in Chapter 18 are a survey of reaction types. Focus on those that are covered by your instructor.

<u>Overview</u>

Based on the number of Exercises, Chapter 18 may very well become your new favorite Chapter, or it may become your new favorite chapter for a more profound reason. Whereas Chapter 17 dealt with the (typically called) area of *inorganic chemistry*, Chapter 18 deals with the chemistry of carbon, that is, the (typically called) area of *organic chemistry*. As this Chapter is a survey of sorts, it might help to provide a few truths about organic chemistry. Simply put, organic chemistry requires less math, and there is really no debating this fact. Simply pick up any organic chemistry book – even those at the graduate student level – and compare it to a physical chemistry book. Even a quick perusal of Chapter 18 clearly indicates an absence of mathematical equations. Further, organic chemists are more likely to spend time in a laboratory, and physical chemists are more likely to spend time solving equations. To oversimplify, there are chemists who care more about *what* happens (organic) and there are chemists who care more about *why* it happens (physical). Both are of equal value, and if you remain in chemistry the choice is yours. (Some care about both, leading the way to physical organic chemistry!)

It is astounding that the simple models of Chapters 6 and 7 can be used to explain so much carbon chemistry. Carbon is simply a very well behaved element that is explained incredibly well by simple sp, sp^2, and sp^3 hybridization. This hybridization also rationalizes the chain-making ability of carbon, probably its most important feature. This allows for a gigantic number of carbon compounds, necessitating the need for systematic naming. Exercises 1 and 3 give you a chance to practice. Additionally, Exercise 3 asks you to consider the concept of chirality, as do Exercises 13 and 14. This initially mysterious concept receives much greater attention in a full organic course than we can provide here, but its importance is vital – literally. See Table 18.5. For Exercise 14 look for carbons that are attached to four different substituents.

18 The Chemistry of Carbon

Exercises 4, 5, 7, and 8 are the quantitative Exercises for this chapter, and they indicate that the reactions, which some might label as "organic", are like any other reaction and are subject to the thermodynamic and kinetic considerations we have seen previously. Exercise 4 is a thermodynamic question for which you must consider the equilibrium constant K from Equation 12.8 for the equilibrium chair $\rightleftharpoons$ boat. The energy difference between the two conformations can be regarded as the free energy change for this calculation. Exercise 5 is a kinetic question on unimolecular (first order) processes (Chapter 15). Exercise 7 is a Hess's Law problem (Chapter 10), and Exercise 8 is another thermodynamic (Chapter 11) Exercise that requires Equation 11.25. Exercise 10 illustrates the reactivity of alkenes. Be certain that your mechanism shows the *movement of electrons*! Draw your first arrow *from* the double bond *towards* the H end of H–Br, not the other way around. The lower E_a for the 2° carbocation, which is more stable than the 1° carbocation, accounts for its rapid formation and influence over the final product.

Organic compounds are typically not acidic. Carboxylic acids are a major exception to this rule, and, to a lesser extent, so are phenol of Exercise 15 and cyclopentadiene of Exercise 16. The relatively large K_a values that these compounds exhibit is attributed in all three cases to a resonance stabilized anion. In Exercise 31 of Chapter 7 and Exercise 27 of Chapter 12, we drew the resonance stabilized carboxylate ion. See the Exercise 15 solution here for the phenolate ion's resonance structures. The aromaticity, which is the ultimate in resonance stabilization, of the cyclopentadienyl ion (Exercise 16) is due to its 5 sp^2 carbons which allow for planarity. (You must also count the number of electrons in the π-system to verify if the $(4n + 2)$ rule is satisfied.) We make special note of these three anions to prepare you for organic chemistry: The stability of the "leaving group", or the end product of a reaction, is commonly used to gauge the feasibility of a reaction. Stable leaving groups imply that a reaction will proceed at a measurable rate.

Another every day event in an organic class is a reaction mechanism. Heretofore, our mechanisms were very simple. See Equations 5.10, 5.12, 5.20, and 5.25, as well as Exercise 37 of Chapter 5. The mechanism shown here for Exercise 11, however, is much closer to the kinds of mechanisms that you will be asked to learn in organic chemistry. Finally, Sections 18.7 and 18.8 are a fearsome twosome for those not familiar with biology *and* organic chemistry. While there may be no math involved, there is a dizzying array of new terms. Plan on several re-reads if this material is assigned to you! Do not forget, however, that you have been prepared for this material. In fact Figure 18.21 is, perhaps, the most cumulative Figure in the text. It contains leaping electrons, enzymes, activation energy, Lewis structures, a mechanism, free energy G, which incorporates H and S, polarity (see anion hole!) and nonpolarity (see greasy pocket!), combined VB/MO theory for the benzene ring, and so on. We guess the gas laws aren't present in Figure 18.21, but almost everything else is! Good luck in your studies and be safe!

1. C_8H_{18} is described by the formula C_nH_{2n+2}, which is the formula for an alkane. The formula for a ring is C_nH_{2n}. There would be two H's left over if we tried to form ring compounds. There are 18 structural isomers for C_8H_{18}. (The modern term for structural isomer is *constitutional isomer*.) This number rapidly grows as we increase the number of carbon atoms. C_9H_{20} has 35 structural isomers, $C_{10}H_{22}$ has 75 structural isomers, and $C_{40}H_{82}$ has 62,491,178,805,831 structural isomers. You can confirm that last one for yourself. See text key for five of C_8H_{18}'s isomers in semistructural form. The skeletal structures of those listed are below:

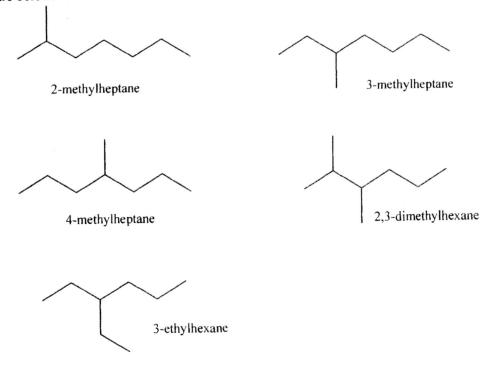

2-methylheptane   3-methylheptane

4-methylheptane   2,3-dimethylhexane

3-ethylhexane

3. A tetrahedral carbon with four different groups attached is a stereocenter and results in a chiral molecule. The structures are:

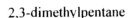

2,3-dimethylpentane   3-methylhexane

5. "Unimolecular" means first order, and Table 15.2 yields $t_{1/2} = \ln2/k$. The rate constant can be obtained from Equation 15.35:

$$k = \frac{1.0 \times 10^{13}}{s} \exp\left(\frac{-10.1\,\text{kcal/mol}}{0.0019872\,\text{kcal/K mol}(298.15\,\text{K})}\right) = 3.95 \times 10^5\,/\text{s}.$$

Thus, $t_{1/2} = 1.8 \times 10^{-6}$ s. Increasing the temperature increases the rate, as usual, because more molecules will have the needed E_a. Lowering the pressure should have no effect until very low pressure is reached. Recall Equation 15.48 in the discussion of the Lindemann mechanism. At this limit unimolecular processes become bimolecular.

7. For Equation 18.3, Table 10.3 and Equation 10.32 give $\Delta H° = 4(99) + 58 - [3(99) + 78 + 103] = -24$ kcal/mol. (Appendix C gives -23.50 kcal/mol). Likewise, for Equation 18.4 use Equation 10.32. See the text answer key for values. So, $\Delta H°_{18.4b} + \Delta H°_{18.4c} = \Delta H°_{18.3}$. This is justification for adding only steps b and c in Exercise 6. The first step of the mechanism,

breaking the Cl_2 bond, should be the rate-limiting step, based on its large, endothermic ΔH° (58.164 kcal/mol).

9. There are six possible structural (constitutional) isomers for C_5H_{10}. See text answer key. Only two of the six exhibit geometrical isomerism, the *cis-* and *trans-* isomers of 2-pentene.

11. The mechanism is below. Note that the final step involves the loss of the H^+ catalyst to a nearby water molecule. This is no problem because mechanisms are *always* after-the-fact rationalizations of experimental facts, and organic chemists use protons very freely when they are needed. Additionally, note that the hydroxyl unit is attached to R′, distinguishing it from the R group of the carboxylic acid.

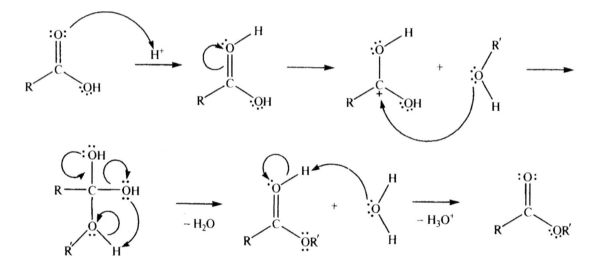

13. There are two molecules that can be drawn with this molecular formula, CH_2ClCH_2F and $CHFClCH_3$. CH_2ClCH_2F can't be chiral because the C's have two H's bonded to them. $CHFClCH_3$, however, has a chiral carbon because it is bonded to four different things: H, F, Cl, and CH_3. Thus, one enantiomeric pair exists.

15. The exact location of the negative charge is impossible to determine, but the resonance structures (see below) indicate that it is delocalized to some extent. Note that the meta positions can't be assigned a negative formal charge, unlike the two ortho and one para positions. To compare, cyclohexanol, whose anion requires that the negative charge be localized solely on the O atom, has a K_a of 1×10^{-18}, whereas phenol has a K_a of 1.3×10^{-10}.

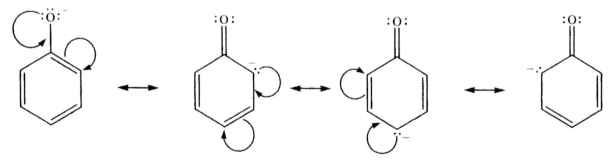

17. See Figures 18.5 and 18.6. The HOMO of ethylene is expected to donate electron density into the LUMO of Br_2 (or Br^+). This donation breaks the Br–Br bond, forming a Br^- ion and

the three-membered C–C–Br$^+$ ring. The Br$^-$ nucleophile then attacks the three-membered ring.

19. The mechanism here is very similar to that of Exercise 11. The only difference is that now there is no generic **R** or **R**'. See the text key for the reaction.